BRITISH CUMACEANS

A NEW SERIES
Synopses of the British Fauna
No. 7

BRITISH CUMACEANS

ARTHROPODA: CRUSTACEA

Keys and Notes for the Identification of the Species

N. S. JONES

Department of Marine Biology,
University of Liverpool,
Port Erin, Isle of Man

1976
Published for
THE LINNEAN SOCIETY OF LONDON
by
ACADEMIC PRESS
LONDON AND NEW YORK

ACADEMIC PRESS INC. (LONDON) LTD
24–28 Oval Road
London, NW1 7DX

U.S. Edition published by
ACADEMIC PRESS INC.
111 Fifth Avenue
New York, New York 10003

Library of Congress Catalog Card Number: 75–46335
ISBN: 0–12–389350 X

Text set in 9/10 pt Monotype Times New Roman, printed by photolithography and bound in Great Britain at The Pitman Press, Bath

Foreword

Cumaceans are small, not very spectacular animals of which a few species may be found burrowing in the sandy and muddy beaches round our shores. Nevertheless they can occur in considerable numbers and are an important part of their fauna as well as being a vital link in many food-chains. Thus this *Synopsis* should be of interest and value to keen naturalists, be they amateur or professional.

The *Synopses* are written by specialists in their respective fields and are designed to fill the gap between popular field-guides and specialist monographs and treatises. *British Cumaceans* is no exception, and the Linnean Society is grateful to Dr Jones for the care he has taken in the preparation of this *Synopsis*.

Spaces have been left throughout the text for the owner's notes and the cover is water-proofed and it is hoped that these features will help the *Synopsis* fulfill its role as a practical field and laboratory guide to this particular group of little crustaceans, the cumaceans.

DORIS M. KERMACK
Synopses Editor,
Linnean Society

A Synopsis of the British Cumaceans

N. S. JONES

*Department of Marine Biology, University of Liverpool,
Port Erin, Isle of Man*

CONTENTS

Introduction

The Cumacea are a very distinctive and uniform order of the Class Crustacea, placed in the Superorder Peracarida of the Subclass Malacostraca. All are easily recognizable by their inflated carapace and pereon followed by a slender pleon ending in a forked tail. All the British species are marine, scarcely penetrating into estuaries. Elsewhere, however, a number of species are known to be adapted to reduced salinity and occur even high up estuaries while some of those found in the Caspian enter the rivers running into it.

Relatively few live between tidemarks and many species are to be found in shallow water offshore, especially in the tropics, but they are also common in the deep sea, with a few recorded from below 7000 m. At the time of writing about 885 species have been described, with at least another 100 collected but awaiting description, while it is likely that several hundred more have yet to be captured. The British fauna, taken in this synopsis to include those recorded from the continental shelf within 20 miles of the British coasts and down to 200 m depth, comprises 41 species. About 20 more may occur on the slope to the west of the British Isles but these have been excluded.

General Structure

A typical cumacean (Fig. 1) has a **carapace** composed of the fused dorsal parts of the **cephalon** (head) and first three thoracic somites which is developed at the sides into overhanging folds enclosing the feeding and branchial appendages. The front of the carapace is produced into a lappet, the **pseudorostral lobe,** on either side, which projects forward and meets its fellow to form the **pseudorostrum**. In the genus *Eudorella* the lobes are reflexed. Below the pseudorostrum the anterior border is excavated to form the **antennal notch** or **sinus** which is defined below by the **anterolateral angle** or **corner**; in many species both are lacking. Behind the pseudorostrum on the dorsum is the bell-shaped **frontal lobe** from which the small **ocular lobe** projects forward, usually carrying the unpaired eyes. The eyelobe normally persists even when eyeless except when the eyes remain separated into two groups as in *Nannastacus*. Behind and to the sides of the frontal lobes are the **branchial regions**.

The last five thoracic somites are together known as the **pereon** (peraeon, pereion) or thorax and each is normally free, although the first and exceptionally the second and third may be coalesced with the carapace, while the third and fourth may be fused together dorsally. Individually each is called a **pereonite** (Fig. 1A). Their **epimera** (coxal plates) are usually feebly developed but the limit between their tergites and sternites is often marked by a longitudinal furrow or ridge.

The abdomen or **pleon** consists of six cylindrical somites, **pleonites,** prolonged by a **telson** in four of the families. In the remainder the telson is coalesced with the sixth pleonite. The anus, protected by a pair of valves, opens on the lower face of the telson or at the end of the sixth pleonite (Fig. 1A and B).

The body is covered with a chitinous epidermis, sometimes strongly calcified. Its surface is often sculptured with grooves, ridges, spines, tubercles or teeth, or carries setae, and has a fine reticulated, pitted or scaled appearance. It is sometimes covered with sand grains.

Between the bases of the first antennae on the underside of the front end of the body are two small chitinous plates, the **epistome,** to the front end of which the first antennae are attached, and behind it the **labrum** or upper lip (Fig. 2A), with the second antennae attached. Behind the labrum is the mouth opening and behind that the **labium** or lower lip, consisting of a chitinous fold attached at its broadened hind end and divided into two lobes which project at the front between the distal ends of the mandibles and first maxillae. The front edge of each lobe carries a row of small setae (Fig. 2C).

The **first antennae (antennulae)** (Fig. 2A) have a peduncle of three segments (articles, joints) and two flagella of which one, the main (outer) flagellum, is usually longer than the other, the accessory (inner) flagellum. The latter is often one-segmented. One or two modified setae, of a jointed appearance and known as **aesthetascs,** may be present at the end of the main flagellum. They are probably sense organs. In the males of some species a brush of sensory filaments is present

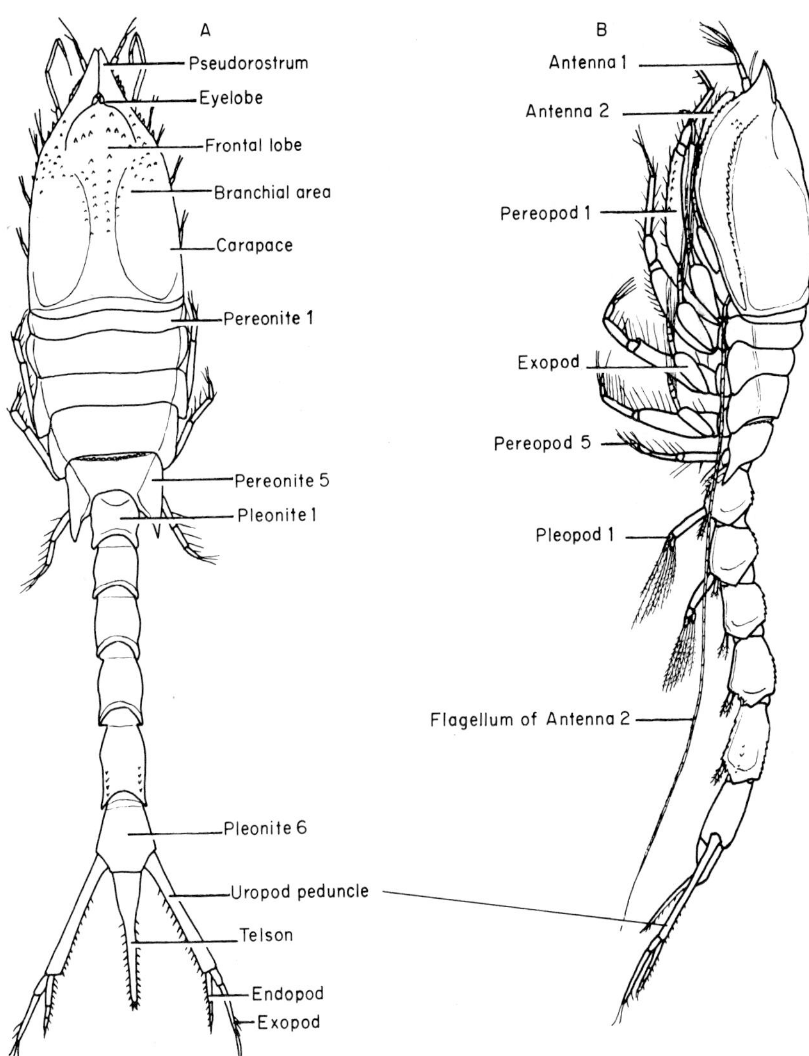

FIG. 1. *Diastylis rathkei:* A, female from above; B, adult male from side (after G. O. Sars, 1900).

on the basal segment of the main flagellum. The **second antennae** are rudimentary in the female, with from one to five segments. In the male (Fig. 2B) there is nearly always a peduncle of five segments and a many-segmented flagellum. The outer sides of the fourth and fifth segments of the peduncle are normally thickly beset with setae. At rest the second antennae are bent backwards between the third and fourth segments of the peduncle and the flagellum is closely applied to the sides of the body. It usually reaches to the end of the body but is sometimes shortened and in *Lamprops* is used as an organ for clasping the female.

The mandibles (Fig. 2D–F) are always without a palp. Each is normally boat-shaped, pointed at either end, with a **molar process** or **pars molaris** jutting inwards from it, and is attached by strong muscles inserted into its hinder part. The front end, the **pars incisiva**, is toothed on the inner edge. A **lacinia mobilis** is present on the left mandible only or also as a rudiment on the right. Between the pars incisiva and the molar process is normally a row of spines. The posterior end of the mandible is sometimes broadened and not pointed. The molar process is usually robust and cylindrical, with a flattened masticating surface, but may be styliform.

The **first maxillae (maxillulae)** (Fig. 3A) lie ventral to the labium and the **second maxillae** (Fig. 3B, C) ventral to the first pair. Both consist of a flattened protopodite bearing endites and the first pair have a backwardly directed palp, usually with one or two filaments, which vibrate in the respiratory chamber and cause a current of water to pass the gills.

The eight pairs of thoracic appendages all normally have seven segments. The first three are modified as **maxillipeds** and the last five as **pereopods**. Each has a short **coxa** fused with the sternite, followed by the **basis**, usually the longest segment, which may carry an exopodite on its proximal part, normally with a peduncle and setiferous flagellum but sometimes a rudiment. Exopods are never present on the first and second maxillipeds nor on the fifth pereopods and the number on the other appendages varies with the species. In the female oostegites are borne on the coxae of the third maxillipeds and the first three pairs of pereopods, lamellar in form and interlocking to form the brood chamber. The remaining segments are the **ischium, merus, carpus, propodus** and **dactyl**.

The first maxillipeds (Fig. 3D) are short and robust but much modified in *Campylaspis* (Fig. 3E) and the coxa carries a greatly developed epipodite, the **branchial apparatus**, made up of a lamellar siphonal part directed forwards, which together with the pseudo-rostral lobes forms a tube, the **siphon**, through which the water leaves the branchial chamber. The backwardly directed branchial part normally carries lamelliform or finger-like lobules or **gills**, often more numerous in the male than in the female.

The second maxillipeds (Fig. 3F) usually have the basis long. The propodus and dactyl are bent in towards one another and may be modified, with large flattened spines (Fig. 3G). In the female there is a flattened plate attached to the hind end of the coxa, the **rudimentary oostegite**, which has a fringe of long slender setae directed backwards into the brood chamber.

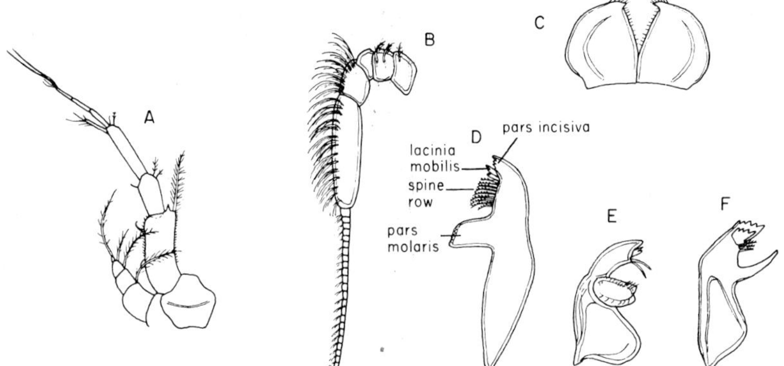

FIG. 2. Antennae and mouthparts: A, female antennae 1 and 2 of *Diastylis rathkei;* B, adult male antenna 2 of *Hemilamprops rosea;* C, labium of *Diastylis rathkei;* D, mandible of *Diastylis rathkei;* E, mandible of *Leucon nasica;* F, mandible of *Campylaspis rubicunda* (all after G. O. Sars, 1900).

The third maxillipeds (Fig. 3H) are large and partly cover the first and second pairs and the mouthparts. They are sometimes similar in structure to the first pereopods but usually differ considerably. Long feathered setae are present on the basis, acting as a screen to filter the water entering the branchial chamber.

The first two pairs of pereopods (Fig. 3I–K) are directed forwards, the first normally reaching beyond the tip of the pseudorostrum. The second are usually shorter than the first and their ischia often fused with the bases. The last three pairs of pereopods are directed backwards and used for burrowing.

Pleopods (Fig. 3L) have been reported only in the female of a single species. In the male there may be one, two, three or five pairs, placed at the hind ends of the pleon sternites, or they may be absent. The peduncle consists of a short coxa and a long basis, with two short rami, the outer usually two-segmented and the inner with a single segment, normally in the adult carrying long plumose setae.

The **uropods** have a one-segmented peduncle and an outer ramus or exopod with two segments, while the inner ramus or endopod may have one, two or three segments. Spines and setae are usually present on the edges of the peduncle and rami and at the ends of the latter. The uropods are used as cleaning organs.

There is always some sexual dimorphism in adults more pronounced in some families than in others, but almost always including better development of the second antennae in the male and the presence of oostegites in the female, frequently the presence of pleopods and of more or better developed exopods on the appendages in the male, and sometimes differences in shape or sculpturing of the integument, besides differences in the development of some sense organs.

The nervous system has a **supraoesophageal mass** and a **chain of 17 pairs of ganglia** joined together by a double nerve trunk. From these, nerves arise to each somite and its appendages. The **eyes**, nearly always situated on a median ocular lobe, are compound, consisting of a small number of **ommatidia** each with a lens-like transparent body beneath the epithelium. In those species living below the photic zone eyes are reduced or more often wholly absent.

The **mouth** opens into a short **oesophagus** leading to the **stomach**, followed by the **mid-gut** which is exceptionally spirally coiled, in turn leading into the **hind-gut** which opens at the **anus** on the telson or sixth pleonite. There are from one to four pairs of **hepatic caeca** which open into the stomach at its junction with the mid-gut.

The sexes are separate. The **ovaries** open on the inner side of the coxae of pereopod 3 and the **testes** on the sternite of pereonite 5.

Further details may be found in Zimmer (1941) and Jones (1963).

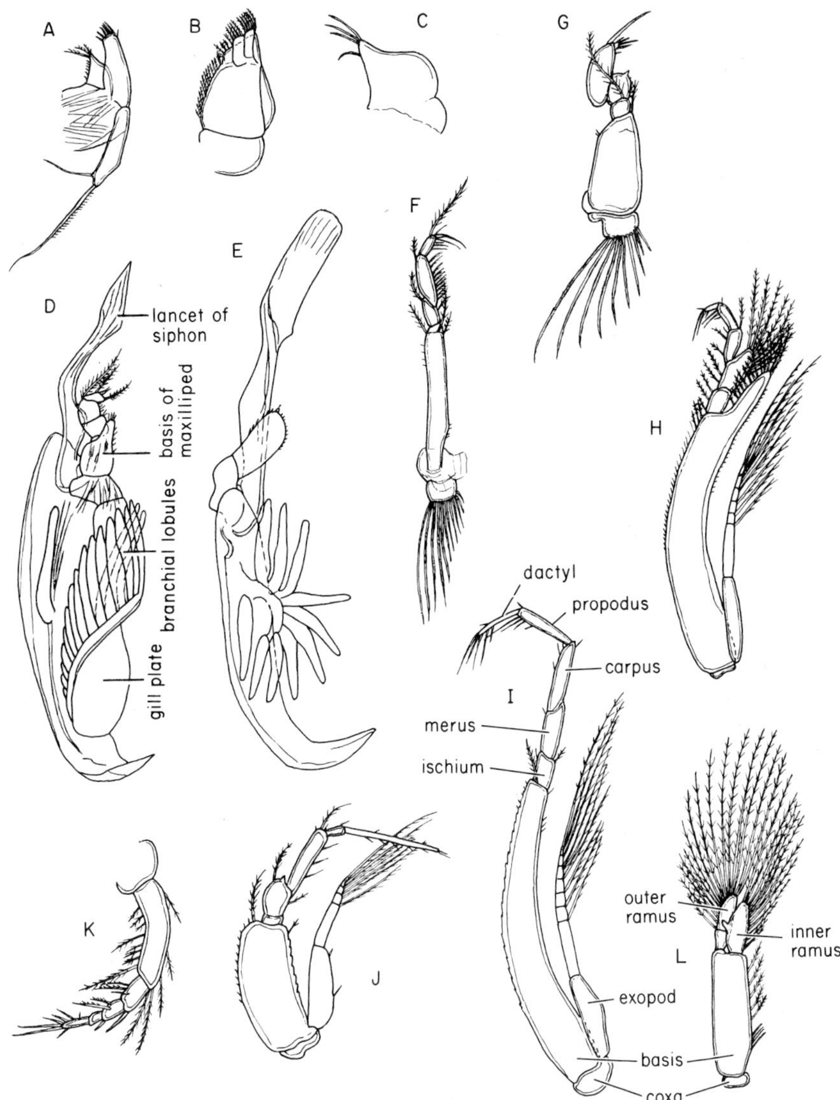

FIG. 3. Appendages: A, maxilla 1 of *Diastylis rathkei*; B, maxilla 2 of *Diastylis rathkei*; C, maxilla 2 of *Campylaspis rubicunda*; D, maxilliped 1 of *D. rathkei*; E, maxilliped 1 of *C. rubicunda*; F, maxilliped 2 of *D. rathkei*; G, maxilliped 2 of *C. rubicunda*; H, maxilliped 3 of *Iphinoe trispinosa*; I, pereopod 1 of *Iphinoe trispinosa*; J, pereopod 2 of *C. rubicunda*; K, pereopod 2 of *I. trispinosa*; L, pleopod of *Bodotria arenosa* (all after G. O. Sars, 1900).

Biology

Life History

Only a few species have been studied and information may be found in Forsman (1938), Krüger (1940), Zimmer (1941) and Corey (1969). The eggs after fertilization are released into the brood chamber in which the larvae are retained until they develop into a stage closely resembling the adult but lacking the last pair of pereopods, the secondary sexual characters, and some of the spines and setae. Brood development takes between one and three months, according to species and water temperature. In different species the number of eggs in the marsupium varies from a few to about 100. After release the larvae moult several times before reaching the subadult stage, when the gonads mature and the secondary sexual characters become apparent though not fully developed, followed by a further moult to the adult, with the full development of a brood pouch in the female and of the second antennal flagellum and pleopods when present in the male. In large species the female may moult at least twice more, leading to a second or third reproductive period, and in some species the subadult males pair and copulate with the adult females. Mating takes place soon after the female has moulted into the adult stage.

The majority of species in temperate shallow waters probably live for a year or less and breed twice each year, but in the deep sea metabolism is probably slowed and species may live much longer, breeding not more than once annually. Little is known about those inhabiting shallow tropical waters. *Diastylis rathkei* breeds once annually and lives for $1\frac{1}{2}$ years while *Cumopsis goodsiri* and *Iphinoe trispinosa* produce two generations a year. In *C. goodsiri* the few survivors of the winter breeding generation reproduce along with the summer generation and the survivors of the summer generation reproduce again in early autumn, the summer generation dying out at the age of 5 months and the winter generation at 12 months. The females of *I. trispinosa* breed once and die soon after the young are released. There is a summer generation between two successive winter generations, the first young released from the winter generation developing rapidly to form the summer breeding generation while those released later in the summer become part of the next winter's generation together with the progeny of the summer generation. The growth rate of the summer generation is rapid and consistent while that of the winter generation is slower and may be completely arrested for a longer or shorter period. *Pseudocuma longicornis* breeds more or less continuously and no distinct generations exist. There is often quite a wide range in size among the adults of any one species.

Feeding

Cumaceans feed on micro-organisms and organic matter from the bottom deposit. Mud-living species filter small particles and those inhabiting sand clean their food off sand grains. In the latter case objects carrying food are picked up by the first pereopods and passed on to the third maxillipeds. The food is then cleaned off by the first and second maxillipeds and passed on to the first and second maxillae and mandibles (Dixon, 1944). The food-manipulating appendages and mouthparts are suitably armed with spines and bristles. In the filter feeders the first and second maxillae together form a pump by means of which water is drawn through their setules which filter off particles of food (Dennell, 1934). In *Campylaspis* and some related genera the mandibles and second maxillipeds are modified as piercing organs and they probably feed on foraminiferans and perhaps small crustaceans.

Behaviour

The majority of species occur on soft deposits and a number on sand where they may show a preference for a certain range of grain size (Foxon, 1936; Pike and Le Sueur, 1958), possibly related to the mesh size of their branchial filtering apparatus or the size of grain which they can manipulate. They normally burrow into the deposit by means of the last four pairs of pereopods, remaining at rest with only the front part of the carapace emerging and sometimes also the uropods, but they make fairly frequent swimming excursions and the adult males of many coastal species may swim up into the plankton, especially at night, and may rise nearly to the surface to form swarms. They may be joined by smaller numbers of females but it is likely that females do not rise as far from the bottom nor remain above it for as long a time. The function of this behaviour is presumably to enable mating to take place or at least for the females to find the males.

Collection and Preservation

Intertidal species may be collected by means of a hand-net or by digging out a portion of sand from the surface and sieving it or shaking it in a dish of seawater and decanting the supernatant liquid through a fine mesh.

Offshore, a runner-dredge such as the Ockelmann (1964) detritus-sledge or that described by Mortensen (1925) will be found most satisfactory. In deep water the epibenthic sledge of Hessler and Sanders (1967), with a closing device, has produced excellent results. On coarser deposits specimens may be collected in a small dredge with a bowed frame and a strong but fine-meshed bag. In each case the resultant catch can be sorted before or after preservation. When alive, specimens may be decanted through a sieve after shaking the deposit in a shallow dish but when preserved it may be necessary to examine the deposit a little at a time through a low power microscope to obtain the smaller specimens.

In shallow water many species may be collected at night by means of a light trap consisting of a coarse-meshed townet suspended just clear of the sea-bed from a boat or jetty, with a low-intensity light fixed in its mouth. After a short period the net is hauled to the surface. It is usual to obtain a preponderance of adult males by this method or they may be the only specimens present.

Cumaceans may be preserved satisfactorily in 70% ethanol with 10% of glycerol added or in Steedman's propylene phenoxytol. Neutral 5% formalin may be used but for long-term preservation ethanol is preferable. Examination and where necessary dissection may be found easier in pure glycerol and more permanent preparations of appendages may be made in glycerin jelly or a mountant of low refractive index. Both sorting and examination may be made easier by staining the sample with rose bengal.

Classification

Order CUMACEA

> Family Bodotriidae
>   Subfamily Vaunthompsoniinae
>   Subfamily Bodotriinae
> Family Leuconidae
> Family Nannastacidae
> Family Ceratocumatidae
> Family Pseudocumatidae
> Family Lampropidae
> Family Diastylidae

Systematic Part

The seven families set out above are separated according to the presence or absence of a separate telson and where it is present its size and armature, the shape of the mandibles, the number of pereopods bearing exopodites, the number of segments in the endopod of the uropod, and the number of pleopods present in the male. Although this classification works well for the British species it is not wholly satisfactory when the world fauna is considered because a few species have been found which partially bridge the gap between families, as for example between Diastylidae and Lampropidae on the one hand and Pseudocumatidae on the other. It is then preferable to refer also to a synopsis of characters such as that in Jones (1963), based on Zimmer (1941), although this now needs some minor modifications following the discovery of further species.

Stebbing (1913) divided the Cumacea into 26 families but some of the characters used are scarcely regarded as differentiating genera in the more usual classification and his scheme has found few adherents.

Bacescu (1972) suggested setting up an eighth family, Archaeocumatidae, based on a species found to have a single pair of pleopods in both male and female, but this species is so similar to *Platysympus* in the Lampropidae that further confirmation is desirable before acceptance of a new family.

No species of the Ceratocumatidae has yet been collected on the shelf but representatives of the other six families occur among the British marine fauna.

More detailed descriptions of the species included in this synopsis may be found in Sars (1900), Stebbing (1913), Fage (1951) or Lomakina (1958) and briefer ones in Jones (1958).

Regrettably the key which follows is based partly on characters of the adult or subadult male only. Although the females and immature males in each family usually have a facies peculiar to that family and different from that of others, this is difficult to quantify and can only be learnt by experience. However, the British species should not be difficult to place in their families.

Key to the Families

(Applicable to the British species only)

1. No independent telson (Fig. 4D) **2**
Telson separated from last pleonite though sometimes small . . . **5**

2. In both sexes the last four pairs of pereopods either without exopods or
with very small rudimentary exopods. ♂ with 5 pairs of pleopods
Family Bodotriidae (p. 14)
In the female at most the last three pairs of pereopods without exopods, in
the male only the last. ♂ with 0, 2 or 5 pairs of pleopods **3**

3. Inner ramus of uropod one-segmented. In ♀ the last three pairs of pereopods
without exopods, in ♂ only the last pair. ♂ without pleopods
Family Nannastacidae (p. 31)
Inner ramus of uropod two-segmented. In ♀ the last two pairs of pereopods
without exopods, in ♂ the last pair. ♂ with 5 or 2 pairs of pleopods **4**

4. Eyes distinct. Mandible of normal shape (Fig. 2D). ♂ with 5 pairs of
pleopods . . . . . Family Bodotriidae—*Vaunthompsonia* (p. 16)
Eyes wanting. Mandibles truncate at base (Fig. 2E). ♂ with 2 pairs of
pleopods . . . . Family Leuconidae (p. 25)

5. ♂ with 5 pairs of pleopods . . . . Family Ceratocumatidae —
♂ with at most 3 pairs of pleopods or none **6**

6. Telson well developed with the anus opening at its base and with 3 or more
apical spines. ♂ with 3 pairs of pleopods or none
Family Lampropidae (p. 42)
Telson well developed with a preanal and a postanal section or small, and
with 2 apical spines or none. ♂ with 2 pairs of pleopods **7**

7. Telson small without apical spines. Endopod of uropod with only one
segment . . . . Family Pseudocumatidae (p. 38)
Telson larger with 2 end spines. Endopod of uropod with 2 or 3 segments
Family Diastylidae (p. 45)

Family BODOTRIIDAE

No free telson. Pleopods with a process on the outer edge of the inner ramus, 5 pairs, occasionally 2 or 3 pairs, in male. Exopodites present on third maxillipeds and first pereopods in the male, or on pereopods 1–4 or sometimes on the first two or three pairs; in the female, besides those on the third maxillipeds, there may be exopodites on the first three or only the first pair of pereopods, occasionally on the first two or first four pairs; in either sex some of the posterior pairs of exopodites may be rudimentary. The mandibles are not broadened at the base. The number of free thoracic somites is often reduced. The family may be divided into two subfamilies according to the number of thoracic exopodites: Vaunthompsoniinae, which in the British fauna include those with well developed or rudimentary exopodites on pereopods 1–3 or 1–4 of the male and 1–3 of the female, and Bodotriinae with exopodites on the first pereopods only in either sex.

Key to the British Species of Bodotriidae

1. Well developed or rudimentary exopods on at least pereopods 1–3 (Fig. 4C, D, F) . . . . Vaunthompsoniinae **2**
Exopodites on first pereopods only (Fig. 5A) . . . . Bodotriinae **5**

2. All the exopodites well developed (Fig. 4C, D)
Vaunthompsonia cristata Bate (p. 16)
The exopodites on pereopods 2 and 3 rudimentary (Fig. 4F) **3**

3. Carapace with two pairs of lateral ridges (Fig. 4E, F)
Cumopsis goodsiri (v. Beneden) (p. 18)
Carapace without lateral ridges **4**

4. Distal segment of inner ramus of uropod with one thin apical spine and several lateral spines; first segment of outer ramus shorter than the second (Fig. 4J) . . . . . . *Cumopsis longipes* (Dohrn) (p. 18)
Distal segment of inner ramus of uropod with only one stout spine at the end and first segment of outer ramus longer than the second (Fig. 4I)
Cumopsis fagei Bacescu (p. 19)

5. The first pereonite visible from above (Fig. 6F) **6**
The first pereonite hidden from above (Fig. 5C) **8**

6. Pleonite 6 with 2 or 4 terminal setae. Adult males with a sternal process on pereonite 2 (Fig. 6D, E, I, J) **7**
Pleonite 6 with 6 terminal setae. Adult males without sternal process (Fig. 6M, N) . . . . *Iphinoe tenella* Sars (p. 22)

7. Carapace with 2–6 teeth dorsally in the female (Fig. 6B), unarmed in the male. 2 terminal setae on pleonite 6. Antenna 1 with a single aesthetasc (Fig. 6C) . . . . *Iphinoe trispinosa* (Goodsir) (p. 22)
Carapace with 8–20 teeth dorsally in either sex (Fig. 6G). 4 terminal setae on pleonite 6. Antenna 1 with two aesthetascs (Fig. 6H)
Iphinoe serrata Norman (p. 22)

8. Carapace without lateral horns. Peduncle of the uropods much longer than the rami (Fig. 5D) **9**
Carapace with lateral horns. Peduncle of uropods much shorter than the rami (Fig. 7B, C) . . . . *Eocuma dollfusi* Calman (p. 24)

9. Carapace with a single carina on either side (Fig. 5A, C) **10**
Carapace with two carinae on either side (Fig. 5E, F)
Bodotria pulchella Sars (p. 20)

10. The inner ramus of the uropod with two segments (Fig. 5B)
Bodotria scorpioides (Montagu) (p. 20)
The inner ramus of the uropod with one segment only (Fig. 5D)
Bodotria arenosa Goodsir (p. 20)

Genus VAUNTHOMPSONIA Bate, 1858

Pereopods 1–4 of the male and 1–3 of the female provided with well developed exopodites. Distal end of the basis of maxilliped 3 little or not at all prolonged. Eyes well developed. There is only one British species.

Vaunthompsonia cristata Bate, 1858

Adult male with carapace smooth, antennal notch shallow (Fig. 4C). Last pleonite projecting backwards between the uropods and with its hind border serrated but these not always apparent (Fig. 4D). Flagellum of antenna 2 short, scarcely reaching beyond pereonite 5. Female with median dorsal crest on the carapace, composed of two parallel rows of small teeth (Fig. 4A, B). The antennal notch is deeper and serrated with a pronounced anterolateral angle below it. The eyelobe is well developed and strongly pigmented. Length of male up to about 5 mm; female up to 6 mm.

In shallow water down to about 40 m on sand or coarse bottoms round the south and west of the British Isles. Also extending to the coast of Morocco and into the Mediterranean and recorded from South Vietnam and South Africa.

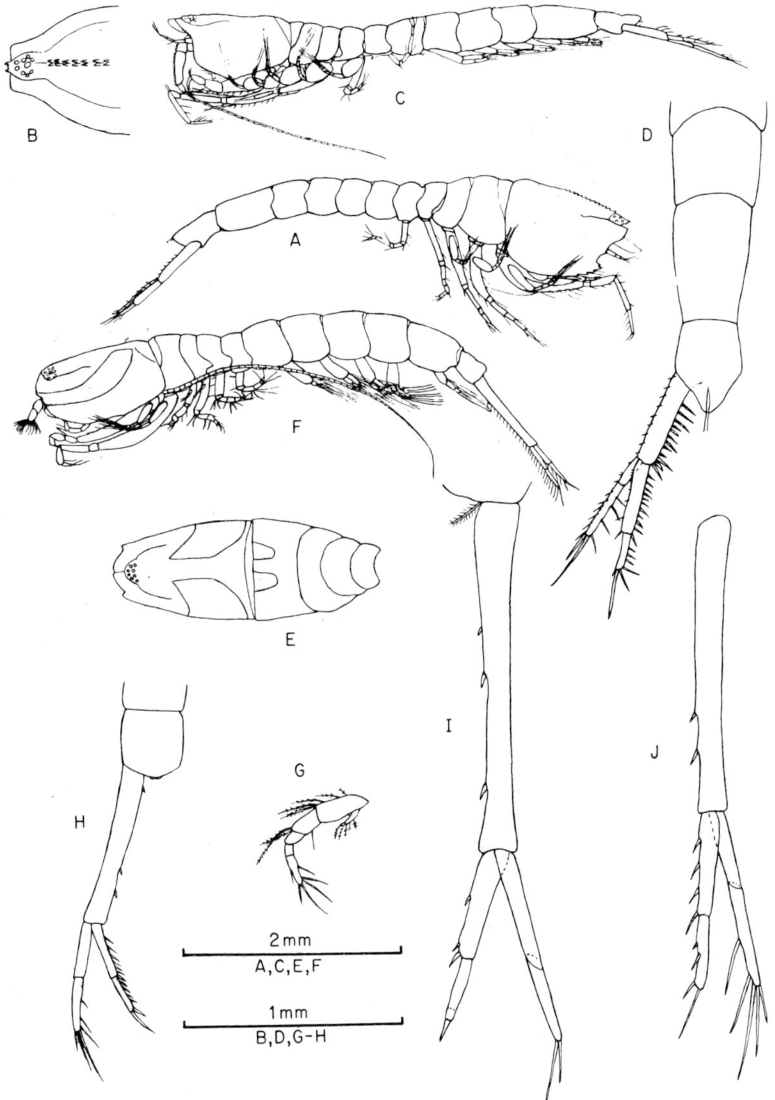

FIG. 4. *Vaunthompsonia cristata:* A, female; B, front of female carapace from above; C, adult male; D, end of pleon and left uropod of male. *Cumopsis goodsiri:* E, female carapace and pereon from above; F, adult male; G, female pereopod 2; H, female left uropod. *Cumopsis fagei:* I, female right uropod. *Cumopsis longipes:* J, female right uropod (I, J after Bacescu, 1956).

Genus CUMOPSIS Sars, 1878

Pereopod 1 with well developed exopodite and pereopods 2 and 3 with rudimentary exopodites in either sex. Male antenna 1 with a tuft of sensory setae at the base of the flagellum. Eye well developed. Two confirmed British species and a third recorded but doubtful.

Cumopsis goodsiri (v. Beneden, 1861)

A pair of lateral ridges is present on either side of the carapace in both sexes but these are sometimes only faintly defined (Fig. 4F). In the female a pair of semicircular folds may sometimes be seen on the dorsum of pereonite 2 (Fig. 4E). In life patches of purplish brown pigment are often present on the carapace and especially on pleonite 5. The flagellum of antenna 2 reaches beyond the end of the pleon. Female uropods as long as the last three pleonites together; the number of spines on the peduncle and the proximal and distal segments of the inner ramus is variable, 0–5, 3–9, and 2–6 respectively having been recorded (Fig. 4H). Length of male up to 5 mm, of female up to 6 mm.

Mainly an intertidal species, burrowing into fine sand from mid-tide downwards on sheltered beaches but also found subtidally down to a depth of a few metres. Generally distributed in the eastern North Atlantic along the coasts of N.W. Europe southwards from Denmark and to Morocco as well as in the Mediterranean and Black Sea. It has also been recorded from the coast of South Vietnam.

Cumopsis longipes (Dohrn, 1869)

C. laevis Sars, 1879

The carapace is smooth, without lateral ridges. There are no semicircular folds on the dorsum of pereonite 2 in the female. The female uropods are distinctly longer than the last three pleonites together and there are 2 or 3 spines on the inner edge of the peduncle and 3 or 4 on the distal segment of the inner ramus (Fig. 4J). In other respects this species closely resembles *C. goodsiri* and there are doubts about its separation. Length of female up to 6 mm.

A littoral species, recorded from the south coasts of the British Isles, from the Bay of Biscay and the Mediterranean. Even if the species should be distinct from *C. goodsiri*, the records from the British Isles must be considered dubious because of the difficulty of identifying specimens of the latter in which the folds on the carapace are indistinct. Bacescu (1956) stated that he had not found *C. longipes* at Roscoff and suggested that previous records were of *C. fagei*. Toulmond (1964) recorded *Cumopsis* sp. from near Roscoff with the characters of both *longipes* and *goodsiri* but Nouvel (1972) records both species from the extreme south-west of France.

Cumopsis fagei Bacescu, 1956

Carapace without lateral folds. Semicircular folds not present on dorsum of pereonite 2. Female uropods as long as the last three pleon somites together; proximal segment of the exopod longer than the distal segment, in contrast to *C. goodsiri* and *C. longipes* in which the distal segment is the longer; the distal segment of the endopod ends in a strong spine but has no spines on the inner edge (Fig. 4I). Length of male up to 6 mm, of female up to 5·8 mm.

Mainly an intertidal species but also occurring in shallow water. It may be found on rather more exposed sandy beaches than those favoured by *C. goodsiri*. It has been recorded from the Channel Islands and from Brittany and Morocco and probably occurs on the south-west coasts of the British Isles.

Genus BODOTRIA Goodsir, 1843

Epidermis strongly calcified. The first pereonite not visible from above, the second long. Only the first pereopods bearing exopodites in either sex. Pereopod 2 with the basis and ischium not distinctly separated. The endopod of the uropod may be either one- or two-segmented with the distal segment always the shorter. There are three British species.

Bodotria scorpioides (Montagu, 1804)

Cuma edwardsii Sars, 1900

Carapace with a longitudinal median carina and a pair of lateral carinae continued to the end of pereonite 5 (Fig. 5A) and visible on the first five pleonites, especially in the male (Fig. 5B). The endopod of the uropod has two segments, the distal much the shorter (Fig. 5B). Length up to 7 mm.

Occurs in some places above low water springs in fine sand and down to 100 m depth but usually much less. Generally distributed round the British Isles and from Norway to France. A forma *mediterranea* has been described from the Mediterranean and the Black Sea, distinguished in the male from forma *typica* by the less distinct carinae on the carapace, the relatively short prolongation of the basis of maxilliped 3 which scarcely reaches beyond the ischium, the shortness of the carpus of pereopod 1, barely as long as the ischium and merus together, and the indistinct separation of the two segments of the endopod of the uropod. This form has also more convincingly been ascribed to *B. arenosa*.

Bodotria arenosa (Goodsir, 1843)

Cuma scorpioides Sars, 1900

Carapace with median and lateral carinae (Fig. 5C) much as in *C. scorpioides*, which it resembles in most respects, the major difference being that the endopod of the uropod has one segment only (Fig. 5D). It is doubtful if it is distinct from *B. scorpioides* forma *mediterranea*. Length up to 7 mm.

Generally inhabits rather coarser sand than *B. scorpioides*, down to about 120 m depth. Probably on suitable substrates all round the British coasts and has been recorded from S.W. Norway to Brittany.

Bodotria plchuella (Sars, 1879)

Carapace with a median and two pairs of lateral carinae, the lower one meeting the upper near the hind end (Fig. 5E, F). Only the median carina is visible on the pleon somites. The basis of pereopod 2 has a number of recurved spines on the upper edge. The endopod of the uropod has two segments (Fig. 5G). Length of male up to 3·2 mm, of female up to 2·5 mm.

Recorded from shallow water down to 70 m in a number of places along the west coast of Britain from the Clyde southwards and also from the Forth. Elsewhere from the Kattegat to the coast of Sénégal and from the Mediterranean.

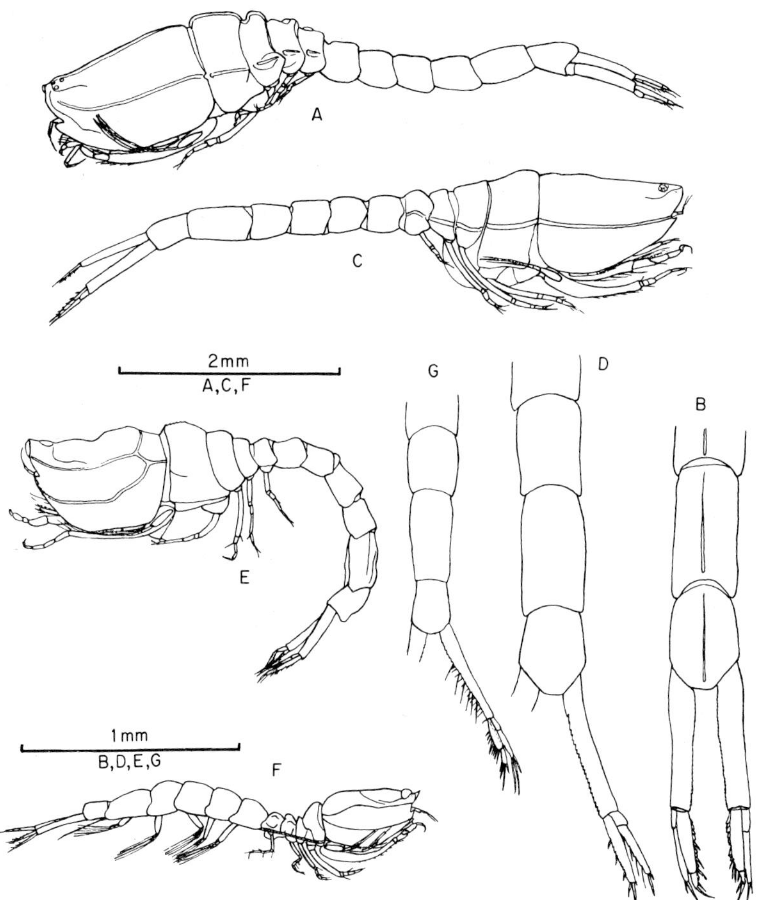

FIG. 5. *Bodotria scorpioides:* A, female; B, end of female pleon and uropods. *Bodotria arenosa:* C, female; D, end of female pleon and right uropod. *Bodotria pulchella:* E, female; F, adult male; G, end of male pleon and right uropod.

Genus IPHINOE Bate, 1856

Body slender and elongated. Carapace compressed laterally and with a mid-dorsal carina. Pseudorostrum prominent. First pereon somite free from carapace. Pereopod 2 with the basis and ischium fused. Endopod of uropod with two segments, the distal usually the longer. Essentially a genus of shallow waters. Three species have been recorded in British waters but it is possible that some of those described from the Mediterranean by Ledoyer (1965) may eventually be found on the south or west coasts.

Iphinoe trispinosa (Goodsir, 1843)

Body very long and slender. Carapace of adult male with dorsal crest unarmed, in female with 2–6, usually 3–4, small teeth in about the middle (Fig. 6B). The pseudorostrum of the female is acutely pointed, blunter in the male. Antenna 1 has one aesthetasc (Fig. 6C). The sternite of pereonite 2 in the adult male has a median tubercle with 6–8 spines on its rounded summit and two lateral tubercles with bifurcated points. The basis of pereopod 2 has a strong bifid spur pointing inwards (Fig. 6D). Pleonite 6 with two short setae posteriorly (Fig. 6E). Length up to about 10 mm.

Occurs on beaches of fine sand and down to about 150 m but usually in shallower depths all round the British coasts and from Norway to the Canary Islands as well as in the Mediterranean.

Iphinoe tenella Sars, 1878

I. gracilis Sars, 1878

Carapace with the dorsal crest in either sex bearing a series of prominent teeth increasing in size towards the front (Fig. 6K) (in some Mediterranean specimens the male has no teeth). Antenna 1 with one aesthetasc (Fig. 6L). In the adult male the sternite of pereonite 2 is scarcely prominent in the mid-line and without tubercles, while the bases of pereopods 2 have no spurs (Fig. 6M). Pleonite 6 has 6 setae at the posterior end (Fig. 6N). Length up to 8 mm.

Found on muddy sand intertidally and in shallow water down to about 30 m. Recorded from the English Channel and from the French coast southwards to the Ivory Coast and in the Mediterranean. It has also been reported from southern India.

Iphinoe serrata Norman, 1867

Body very long and compressed. Carapace in either sex with the mid-dorsal crest bearing 8–20 teeth, starting just behind the eyelobe and continuing in diminishing size to near the hind end (Fig. 6G). The pseudorostrum is very long. Antenna 1 with two aesthetascs (Fig. 6H). The sternite of pereonite 2 in the adult male is armed with three tubercles, each with a bifid extremity, of which the median is usually the largest (Fig. 6I). There are no spurs on the bases of pereopods 2. Pleonite 6 has 4 setae at its hind end (Fig. 6J). Length up to 12 mm.

Found on muddy sand, normally in somewhat deeper water, 30–150 m, than most others in the genus. Recorded from the Shetland Islands, the Irish Sea, south-west Ireland, and from the Bay of Biscay and the Mediterranean.

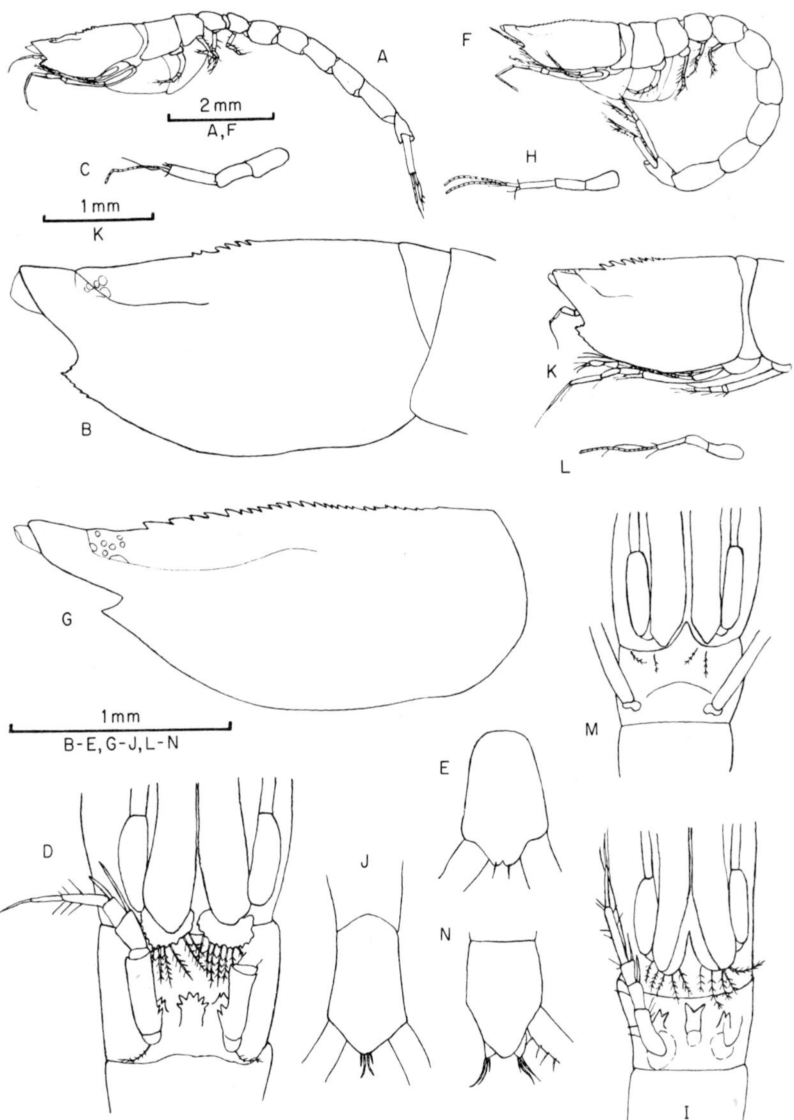

FIG. 6. *Iphinoe trispinosa*: A, female; B, female carapace; C, antenna 1; D, male pereonites 4 and 5 from below; E, female pleonite 6. *Iphinoe serrata*: F, female; G, female carapace; H, antenna 1; I, male pereonites 4 and 5 from below; J, female pleonite 6. *Iphinoe tenella*: K, female carapace; L, antenna 1; M, male pereonites 4 and 5 from below; N, female pleonite 6.

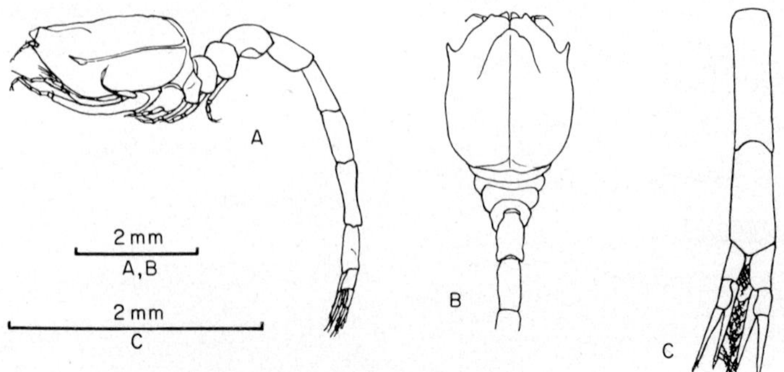

FIG. 7. *Eocuma dollfusi:* A, female; B, female carapace and pereon from above; C, end of female pleon and uropods.

Genus EOCUMA Marcusen, 1894

Epidermis strongly calcified and brittle. Carapace with lateral horns, at least in the female. First pereonite and sometimes the second firmly united with the carapace. The basis of pereopod 1 prolonged at its distal inner edge. Pereopod 2 with basis and ischium fused. Uropods with the peduncle much shorter than the rami, of which the inner has one segment only. One species found in the Channel Islands which may reach the south coast of England.

Eocuma dollfusi Calman, 1907

Carapace with one pair of lateral horns each having an acute point directed forwards, behind which a well-marked keel runs backwards to the hind end (Fig. 7A, B). Body smooth, without hairs or spines. Uropods with the rami nearly three times as long as the peduncle (Fig. 7C). Length up to 7 mm.

Found in sand intertidally in Jersey and from the Normandy coast to Morocco.

Family LEUCONIDAE

No free telson. Pleopods without an external process on the inner ramus, 2 pairs, rarely 1 or 0, in male. Exopodites present on the third maxillipeds and on the first four pairs of pereopods, rarely the first two pairs, in the male; in the female on the first three pairs, rarely the first two pairs. The mandibles are broadened at the base. The endopod of the uropod is two-segmented or rarely one-segmented. All the pereon somites are visible from above.

Key to the British Species of Leuconidae

1. Pseudorostrum distinct with the efferent orifice at the front (Fig. 8A, B)
                Leucon nasica (Kröyer) (p. 26)
 Carapace truncate anteriorly without prominent pseudorostrum, which is reflexed; efferent orifice dorsal (Fig. 9A) **2**

2. Inner ramus of the uropod shorter than the outer (Fig. 9J)
             Eudorellopsis deformis (Kröyer) (p. 30)
 Inner ramus of the uropod longer than the outer (Fig. 9D) . . . **3**

3. Tooth below sinus at front of carapace small, not projecting beyond upper teeth (Fig. 9A, B) . . . . . . *Eudorella truncatula* (Bate) (p. 28)
 Tooth below sinus large, projecting beyond the upper teeth (Fig. 9E, F)
             Eudorella emarginata (Kröyer) (p. 28)

Genus LEUCON Kröyer, 1846

Pseudorostrum prominent with efferent orifice at front. Carapace with longitudinal mid-dorsal serrate crest in female, often wanting in male. Peduncle of antenna 1 without a marked geniculation. Antenna 2 of female with a well defined distal segment. Only one species has been recorded from the shelf round the British Isles but one or two others might yet be found in the north.

Leucon nasica (Kröyer, 1841)

Body slender and elongate. Pseudorostrum a little upturned in the female, nearly horizontal in the male, obliquely truncate and setiferous in the female with a deep antennal notch and acute anterolateral angle below; in the male the lobes are shorter, almost transversely truncate, with the antennal notch scarcely excavated and the anterolateral angle rounded and barely prominent. In the female the mid-dorsal crest is serrate from behind the pseudorostrum to beyond the middle and after an interval again serrate to the hind margin (Fig. 8A); in the adult male there are no serrations (Fig. 8B). Pereopod 3 of adult male with two conspicuous unequal ensiform setae on the ischium extending twice as far as the rest of the appendage (discounting the setae or spines on the carpus, propodus and dactyl) (Fig. 8C). The peduncle of the uropods is as long as the rami; the distal segment of the endopod is less than half as long as the proximal (Fig. 8D). Length of male up to 10 mm, of female up to 12 mm.

An Arctic-Boreal species, probably widespread on muddy bottoms round the British Isles except on southern coasts. Recorded elsewhere on both sides of the Atlantic in colder water and extending into the Arctic and also from Alaska, usually in moderate depths between 10 and 100 m but found down to more than 500 m.

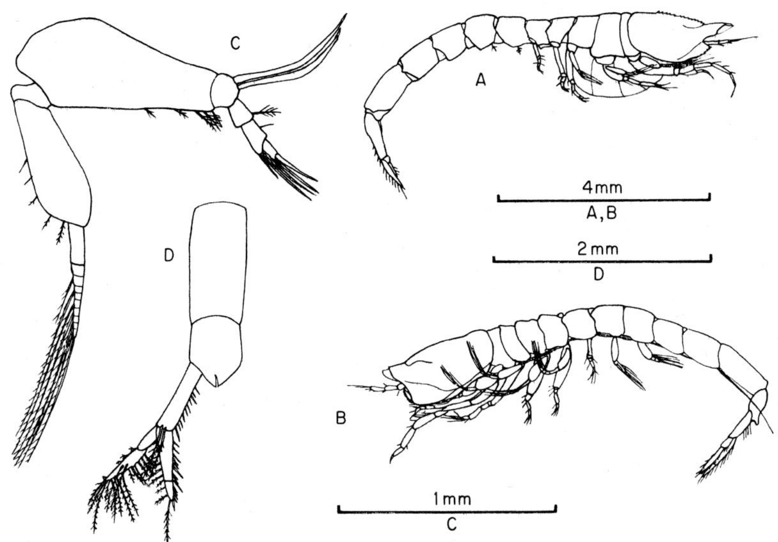

FIG. 8. *Leucon nasica:* A, female; B, adult male; C, male pereopod 3; D, end of female pleon and left uropod.

Genus EUDORELLA Norman, 1867

Carapace short, subtruncate, without dorsal serrations, pseudorostrum reflexed with efferent orifice dorsal. Antenna 1 bent (geniculate) between the second and third segments. Distal segment of antenna 2 in female ill-defined. Uropods with the exopod shorter than the endopod. There are two British species.

Eudorella emarginata (Kröyer, 1846)

Body slender, elongate, finely hairy. Front fringed with short setae, smooth except for some serrations above a semicircular excavation which is bounded below by a slightly upturned acute tooth which projects beyond the upper serrations; the excavation is much smaller in the male than in the female. The lower margin is serrate for some distance back (Fig. 9E–G). A pair of long setae project from the hind end of pleonite 5. Peduncle of uropod a little longer than the exopod, about equal to the proximal segment of the endopod, of which the distal segment, including its terminal spine, is much less than half as long (Fig. 9H). Length up to 12 mm.

An Arctic-Boreal species found on muddy sand and mud bottoms round the British Isles except on southern coasts. Elsewhere on both sides of the Atlantic in colder water and extending into the Arctic, from about 10 to 2000 m depth but usually on the shelf.

Eudorella truncatula (Bate, 1856)

Body slender, covered with short hairs. Front of carapace in the female with a few stiff setae on either side on the rounded upper front corner, smoothly truncate in front down to the denticulate, slightly prominent upper edge of a slight excavation, the middle of which is occupied by a prominence with two or three small downward pointing teeth, while its lower edge is formed by an upward pointing tooth followed by some serrations on the lower edge; in the male the excavation is replaced by a smooth convexity with only a few teeth on the lower front edge (Fig. 9A–C). There are no long setae on the hind border of pleonite 5. Peduncle of uropods a little longer than the exopod, which is a little longer in the female or a little shorter in the adult male than the first segment of the endopod. The distal segment, including its terminal spine, is more than half as long as the proximal segment of the endopod in the female, less than half in the male (Fig. 9D). Length up to 5 mm.

A Mediterranean-Boreal species, found on muddy sand substrates all round the British Isles, usually in less than 50 m depth. Found on both sides of the North Atlantic and in the Mediterranean but does not extend into the Arctic. Recorded between about 10 and 550 m.

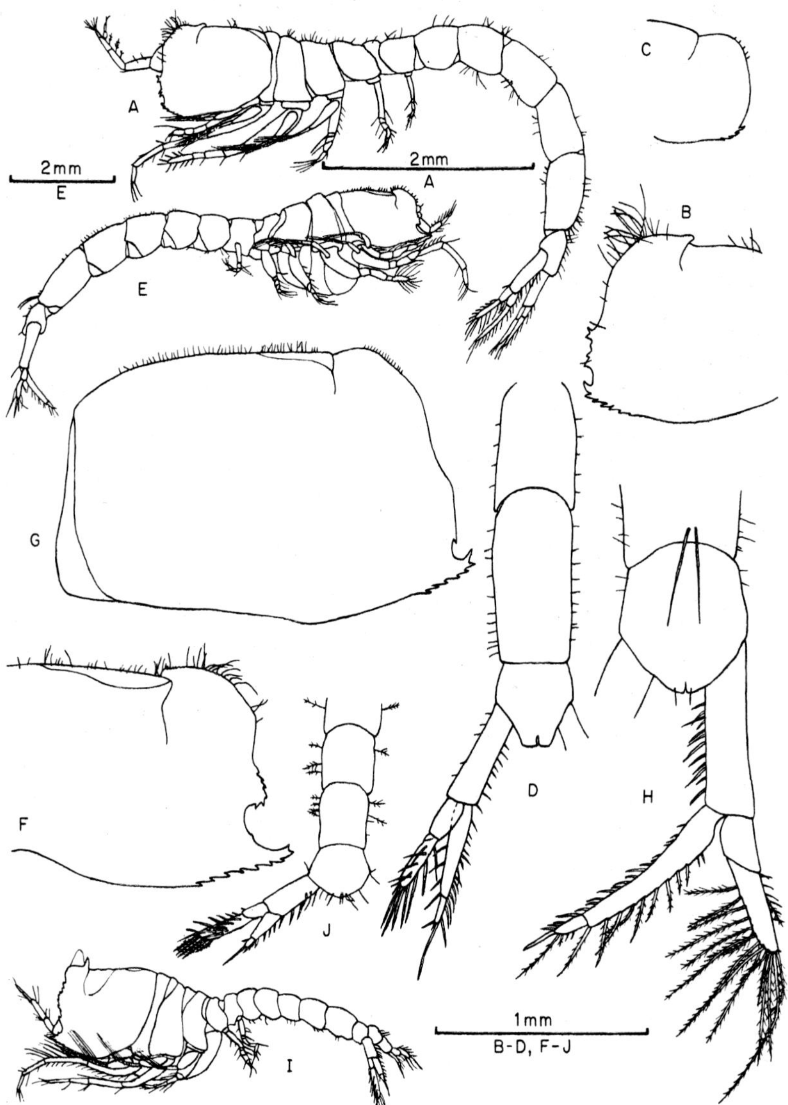

FIG. 9. *Eudorella truncatula:* A, female; B, front of female carapace; C, front of male carapace; D, end of female pleon and uropods. *Eudorella emarginata:* E, female; F, front of female carapace; G, male carapace; H, end of female pleon and right uropod. *Eudorellopsis deformis:* I, female; J, end of female pleon and left uropod.

Genus EUDORELLOPSIS Sars, 1883

Resembling *Eudorella* but body shorter. Antenna 1 geniculate between the first and second segments. Uropods with the exopod longer than the endopod. Only one British species.

Eudorellopsis deformis (Kröyer, 1846)

Pseudorostral lobes with a sharp process on the dorsal side behind the efferent orifice, front steeply truncate and serrated throughout in the female with broad flattened teeth, less so in the adult male. In the female the anterolateral corner is prominent and the lower margin is serrated for some distance back (Fig. 9I). Peduncle of the uropod much shorter than the exopod, which is much longer than the endopod; the distal segment of the endopod is not more than a third as long as the proximal segment, excluding its small terminal spines (Fig. 9J). Length up to 5 mm.

A Boreal species, found off the north and west coasts of the British Isles on sandy bottoms mainly in shallow water. Recorded elsewhere from Norway southwards to the Irish Sea and from Greenland to Long Island in the Atlantic and also from the Japan and Okhotsk Seas in the Pacific, depth range 15–271 m.

Family NANNASTACIDAE

No free telson. No pleopods. Exopodites usually present on the third maxillipeds and first four pairs of pereopods in the male, rarely on the first two or three pairs; in the female usually on the third maxillipeds but occasionally absent from these appendages, and on the first two pairs of pereopods, rarely the first three pairs or entirely lacking. Mandible normal in shape or with the base widened and the molar process styliform. The endopod of the uropod has a single segment.

Key to the British species of Nannastacidae

1. Molar process of mandible styliform, pointed (Fig. 11B). Anterolateral angle of carapace absent or only slightly prominent (Fig. 11A, I) . . . 2
Molar process of mandibles thick and truncate. Carapace of female with well developed anterolateral angle (Fig. 10A, F) 6

2. Carapace smooth without folds or carinae or lateral depressions . . 3
Carapace with lateral carinae or depressions 4

3. Dactyl of pereopod 2 distinctly longer than the carpus and propodus together (Fig. 11D) . . . . *Campylaspis rubicunda* (Lilljeborg) (p. 35)
Dactyl of pereopod 2 not longer than carpus and propodus together (Fig. 11G)
Campylaspis glabra Sars (p. 35)

4. Carapace with several curved folds on either side (Fig. 11I)
Campylaspis costata Sars (p. 35)
Carapace with a longitudinal depression on either side 5

5. Carapace with longitudinal depressions each divided into two unequal parts (Fig. 11J) . . . . *Campylaspis legendrei* Fage (p. 36)
Carapace with depressions undivided (Fig. 11K)
Campylaspis sulcata Sars (p. 36)

6. A single median ocular group (Fig. 10F) . *Cumella pygmaea* Sars (p. 34)
Two ocular groups widely separated (Fig. 10B, C) 7

7. Uropods about two-thirds as long as the last two pleonites together (Fig. 10E)
Nannastacus brevicaudatus Calman (p. 32)
Uropods longer than the last two pleonites together (Fig. 10D)
Nannastacus unguiculatus (Bate) (p. 32)

Genus NANNASTACUS Bate, 1865

Anterolateral angles of carapace prominent, especially in the female. Eyes, when present, divided into two separate groups. Maxilliped 2 with 6 segments. Endopod of the uropods longer than the exopod. Exopodites may be lacking from maxilliped 3 and pereopods 1 and 2 but are usually present. A large genus mainly from warm water, with two British species.

Nannastacus unguiculatus (Bate, 1859)

Body short and broad. Integument calcified, with many small tubercles and scattered hairs in the female, which is often covered with particles of the deposit. Female with pseudorostral lobes fairly short, upturned, their edges broadly serrate, meeting below the branchial orifice. The anterolateral margins are concave and acutely projecting forwards on either side, with the lower margins coarsely serrated. There are a number of rows of spines on the carapace and these are large and flattened behind the branchial region, while some similar spines occur at the sides of the outstanding sideplates of the pereon somites. Many of these spines, however, may be wholly broken off (Fig. 10A). There are two rows of smaller spines on the pereon and pleon somites. The eyes are wide apart, each with several lenses (Fig. 10C). Uropods longer than the last two pleonites together, the peduncle less than half as long as the endopod, which is about three times as long as the exopod (Fig. 10D). Male (Fig. 10B) with scattered hairs. Pseudorostral lobes shorter and thicker, less upturned, anterolateral corners quadrate. Carapace without flattened spines. Length up to 2 mm.

A Mediterranean-Boreal species, found on coarser substrates in shallow water found the British Isles and from north of Shetland to the Bay of Biscay and also in the Mediterranean, down to about 40 m depth.

Nannastacus brevicaudatus Calma

Near to *N. unguiculatus* but pseudorostral lobes very short, not rising above eyes in lateral view, anterolateral corners less projecting and finely serrated below. Surface of carapace with or without flattened spines in female. Uropods about two-thirds as long as the last two pleonites together, the peduncle about two-thirds as long as the endopod (Fig. 10E). Length up to 2 mm.

Recorded only from the west of Ireland in very shallow water and from the coast of Morocco.

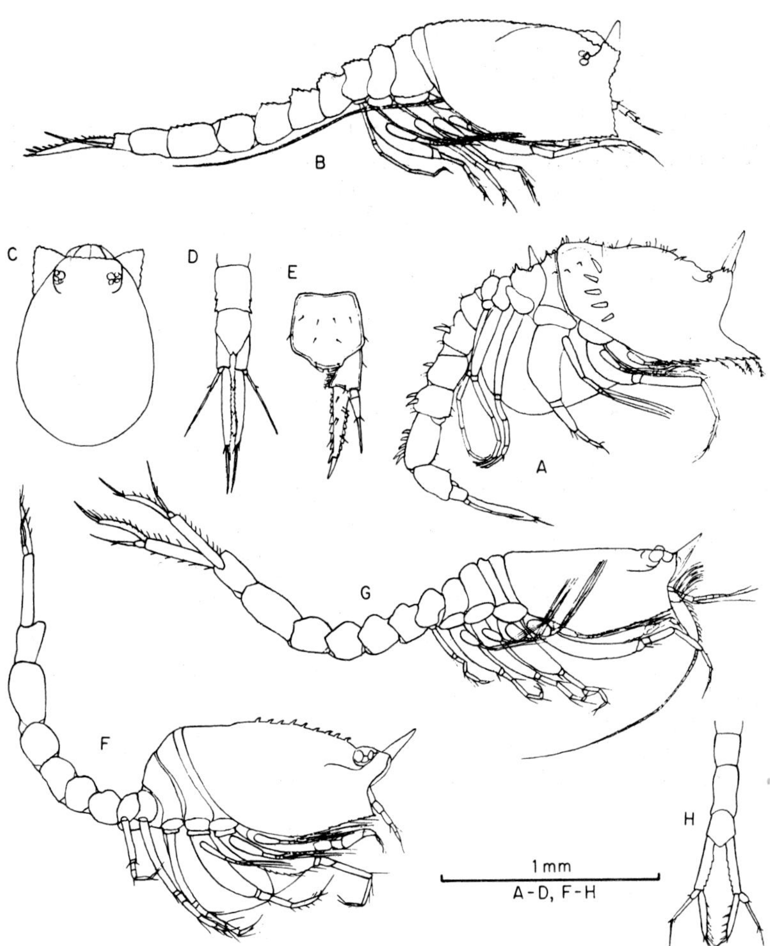

FIG. 10. *Nannastacus unguiculatus:* A, female; B, adult male; C, male carapace from above; D, end of female pleon and uropods. *Nannastacus brevicaudatus:* E, female pleonite 6 and right uropod. *Cumella pygmaea:* F, female; G, adult male; H, end of female pleon and uropods (E, after Calman, 1905).

Genus CUMELLA Sars, 1865

No eyes or a single median ocular group. Maxilliped 2 with 6 segments. Peduncle of the uropods usually longer than the rami and the inner ramus usually longer than the outer. A single British species.

Cumella pygmaea Sars, 1865

Female with pseudorostrum short and a well excavated antennal notch with acute anterolateral angle below. The carapace is compressed with the dorsal edge arched and with a carina bearing 8–12 forward pointing teeth in the mid-line. Eye conspicuous (Fig. 10F). Peduncle of uropods serrated on inner edge, considerably longer than the endopod which is much longer than the narrower exopod (Fig. 10H). In the adult male the pseudorostrum is even shorter, the dorsal line is nearly straight and without teeth (Fig. 10G). The flagellum of antenna 2 reaches only to the third pleon somite. Length up to 3 mm.

Found in shallow water all round the British coasts, mainly in coarse sands and shelly substrates. Elsewhere recorded from Norway to the Bay of Biscay and the Mediterranean and Black Sea, down to 120 m but usually in less than 50 m depth.

Genus CAMPYLASPIS Sars, 1865

Hind end of carapace in female often overhangs the anterior pereon somites. Molar process of mandibles styliform and their base widened. Maxilliped 2 with 6 segments, the propodus articulated at a rightangle to the carpus and ending in a broad seta, the dactyl very short, provided with strong distal diverging spines. Pereopod 1 with the ischium not specially long. A very large genus of which five species have been recorded from the shelf in British waters.

Campylaspis rubicunda (Lilljeborg, 1855)

Carapace large and inflated, dorsal edge arched in the female, less so in the male, smooth, without antennal notch or anterolateral angle. Pereonites 1 and 2 raised dorsally to form forward curving ridges, each with a median point. Eye distinct, fairly prominent (Fig. 11A). Maxilliped 2 with four strong spines on the dactyl (Fig. 11C). Pereopod 2 with the dactyl slender, longer than the carpus and propodus combined (Fig. 11D). Colour mainly red during life. Length up to 6 mm.

An Arctic-Boreal form of the shelf and slope, it does not seem to have been recorded from the southern parts of Britain but has been found on muddy sand in the Irish Sea and northwards and on both sides of the North Atlantic, extending into the Arctic, as well as in the north-western Pacific, in depths from about 10 to 2200 m.

Campylaspis glabra Sars, 1879

Similar to *C. rubicunda* but smaller as adult and colour when alive whitish sometimes with a tinge of pink (Fig. 11F). Pereonites 1 and 2 less produced dorsally. Dactyl of maxilliped 2 with two or three spines. Dactyl of pereopod 2 shorter than carpus and propodus combined (Fig. 11G). Length up to 4 mm.

A Mediterranean-Boreal species of the shelf and slope it probably occurs on muddy sand or mud substrates all round the coasts of Britain. Elsewhere it has been recorded from northern Norway to North-West Africa, from the Mediterranean and from the North-West Atlantic, and also doubtfully from South Vietnam, between about 5 and 3000 m depth.

Campylaspis costata Sars, 1865

Carapace with dorsal outline arched in the female, less so in the male, sides sculptured with three pairs of oblique folds, the hindermost bifurcating so that a fourth fold nearly reaches the hind margin and then turns forward to rejoin the third near the mid-line. Eye distinct, with lenses. Pereonites 1 and 2 produced dorsally, each with a median point (Fig. 11I). Pereonite 5 and pleonites 1–4 each with a pair of dorsolateral tubercles. Dactyl of maxilliped 2 with three spines, the middle one small. Dactyl of pereopod 2 not longer than the carpus and propodus combined. Colour in life reddish-brown. Length up to 6·5 mm.

A Boreal species found on muddy sand and mud from the Irish Sea and Helgoland to northern Norway in depths of about 40 to 500 m but mainly on the shelf and recorded once from 1500 m in the North-West Atlantic.

Campylaspis legendrei Fage, 1951

Carapace with integument finely rugose, marked on each side by a shallow longitudinal depression divided into two unequal parts by a transverse ridge. Eye fairly prominent, with lenses. There is a slightly excavated antennal notch and a rounded anterolateral angle. The first and second pereonites are raised dorsally (Fig. 11J). Dactyl of maxilliped 2 with four spines. Dactyl of pereopod 2 about as long as carpus and propodus combined. Length up to 4 mm.

Recorded at present only from the Irish Sea on coarse sand and shell or gravel in depths between 25 and 60 m and from littoral sand at Concarneau, while a subspecies has been described from about 175 m depth off the coast of Mauritania.

Campylaspis sulcata Sars, 1870

Carapace well arched dorsally in female, dorsal outline with slight undulations, lateral faces with a fairly deep excavation between two carinae which extend obliquely from near the mid-line in the hinder part of the carapace to the base of the pseudorostrum. The eye is fairly prominent, with lenses. The antennal notch is faintly indicated but there is no anterolateral angle. The first two pereonites are elevated dorsally (Fig. 11K). Dactyl of maxilliped 2 with four spines. Dactyl of pereopod 2 only slightly tapered and longer than the carpus and propodus combined (Fig. 11L). Length up to 5 mm.

A Mediterranean-Boreal species of the deeper shelf and slope, it has been recorded from the west coast of Ireland and from Norway to the Mediterranean in depths of about 130 to 650 m.

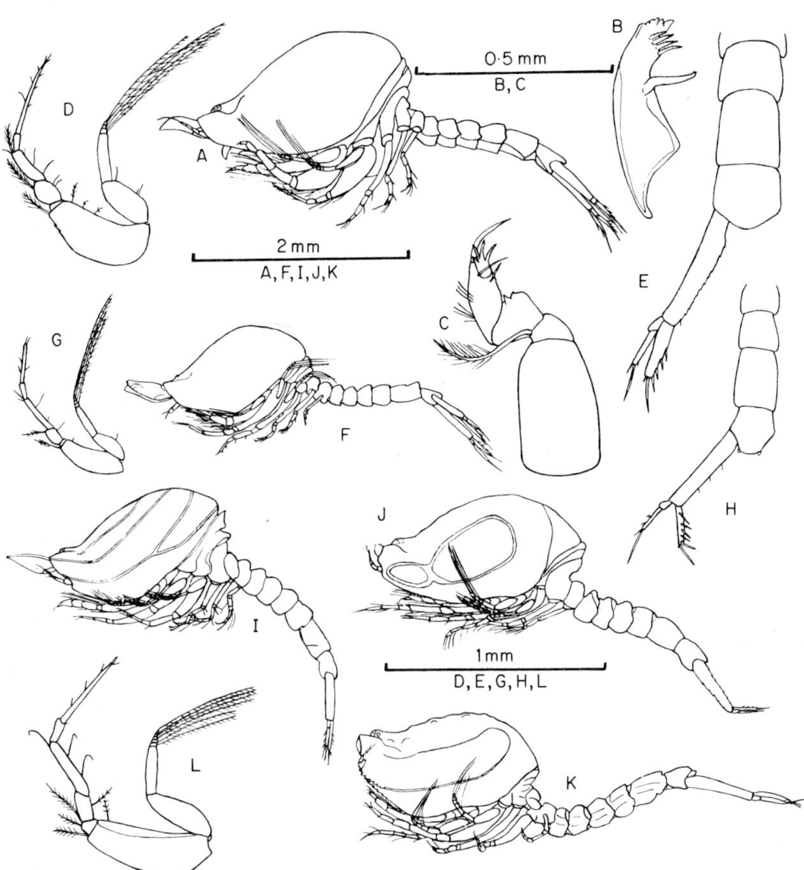

FIG. 11. *Campylaspis rubicunda:* A, female; B, mandible; C, maxilliped 2; D, pereopod 2; E, end of female pleon and left uropod. *Campylaspis glabra:* F, female; G, pereopod 2; H, end of female pleon and left uropod. *Campylaspis costata:* I, female. *Campylaspis legendrei:* J, female. *Campylaspis sulcata:* K, female; L, peroepod 2.

Family PSEUDOCUMATIDAE

Telson present but small. Pleopods without external process of inner ramus, two somewhat rudimentary pairs in the male or occasionally a single pair. Exopodites present on third maxillipeds and first four pairs of pereopods in the male; in the female on the third maxillipeds and well developed on the first two pairs of pereopods, rudimentary on the third and fourth pairs. The inner ramus of the uropod with a single segment.

Key to the British Species of Pseudocumatidae

1. The first pereopod shortened and peculiarly formed (Fig. 12K)
                                        Petalosarsia declivis (Sars) (p. 41)
   The first pereopod not shortened, of normal shape **2**

2. Anterolateral angle of the carapace armed with 3 teeth (Fig. 12F). Female telson broader than long, truncate (Fig. 12G)
                                        Pseudocuma similis Sars (p. 40)
   Anterolateral angle without teeth (Fig. 12B). Female telson nearly semi-circular (Fig. 12D) **3**

3. Antennal flagellum of male not reaching hind end of pereon. Peduncle of uropod with only one inner spine; the inner ramus with few (5–7) spines
                                        Pseudocuma gilsoni Bacescu (p. 40)
   Antennal flagellum of male reaching at least to pleonite 5. Peduncle of uropod with 4 inner plumose setae; the inner ramus with about 12 spines
                                        Pseudocuma longicornis (Bate) (p. 40)

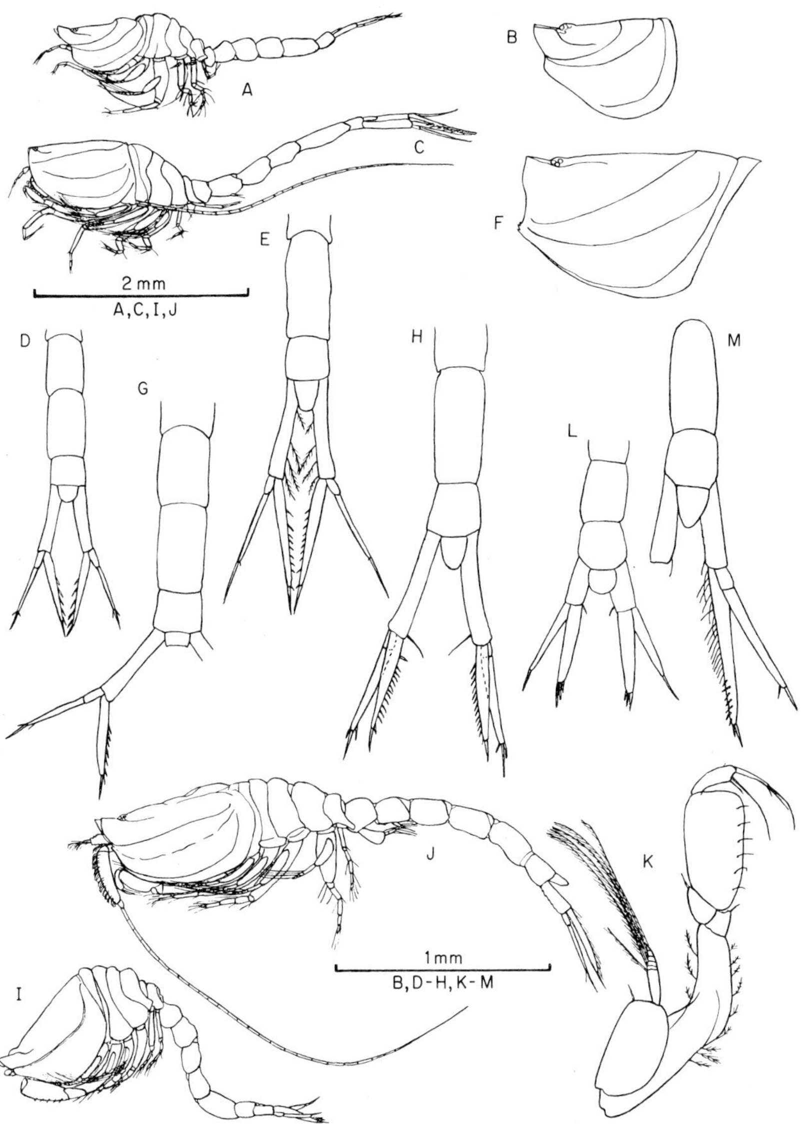

FIG. 12. *Pseudocuma longicornis:* A, female; B, female carapace; C, adult male; D, end of female pleon and uropods; E, end of male pleon and uropods. *Pseudocuma similis:* F, female carapace; G, end of female pleon and left uropod; H, end of male pleon and uropods. *Petalosarsia declivis:* I, female; J, adult male; K, pereopod 1; L, end of female pleon and uropods; M, end of male pleon and right uropod.

Genus PSEUDOCUMA Sars, 1865

Pereon somites without carinae or protuberances. Telson semicircular or bluntly semiovate. Eyes well developed. Second antenna of female very small, one-segmented. Peduncle of uropods not very short. There are two British species and one other of doubtful validity.

Pseudocuma longicornis (Bate, 1858)

P. cercaria Sars, 1900

Carapace with three oblique folds on either side. Anterolateral angle unarmed. Pseudorostrum prominent and acute in female, less so in male (Fig. 12A–C)· Telson nearly semicircular in female, rather longer in male (Fig. 12D, E). Antenna 2 of male with flagellum reaching at least to pleonite 5. Peduncle of uropod equal in length to exopod in female, shorter in male, endopod much longer, with about 12 spines (Fig. 12D, E). Length up to 4 mm.

A widespread species, found in shallow water all round the British Isles and from northern Norway to the Mediterranean and Black Sea, depth 0–130 m. Also recorded from South Africa and from South Vietnam. Bacescu (1950) has described three subspecies but Foxon (1936) regarded males with short antennae found at Millport as sexually mature but not yet fully grown and the systematics of the group would repay further investigation.

Pseudocuma gilsoni Bacescu, 1950

Very similar to *P. longicornis* but male antenna 1 has a brush of aesthetascs on the distal segment of the peduncle and the flagellum of antenna 2 does not reach beyond the end of the pereon, while there are only 5–7 spines on the endopod of the uropod. Length up to 3 mm.

Recorded from the coast of Belgium and from Liverpool Bay in shallow water. It seems possible that this is a neotenous form of the male of *P. longicornis* but further study is required.

Pseudocuma similis Sars, 1900

Similar to *P. longicornis* but less slender and may be somewhat larger. Anterolateral angle with three small teeth (Fig. 12F). Telson rather broader than long, truncate in the female (Fig. 12G). Peduncle of uropod longer than the nearly equal rami (Fig. 12G, H). Length up to 5·5 mm.

A Boreal species, found from Norway to the British Isles and the Bay of Biscay, depth about 10 to 360 m. There is a single record from the Mediterranean.

Genus PETALOSARSIA Stebbing, 1893

Generally similar to *Pseudocuma* but eye poorly developed, antenna 2 of female distinctly two-segmented, pereopod 1 with ischium and merus united to form a cup for the large lamellar carpus, over the broad distal end of which the slender propodus and dactyl can be folded to form a subchelate arrangement. A single species recorded from British waters.

Petalosarsia declivis (Sars, 1865)

Not unlike *Pseudocuma longicornis* but the female carapace has only a single dorsolateral carina on either side, running back to the hind end. In the male there is a second pair of carinae below that present in the female (Fig. 12I, J). The second segment of antenna 1 is rather broad. The first pereopods differ as described for the genus (Fig. 12K). The peduncle of the uropods is short, little more than half as long as the endopod, which is rather longer than the exopod (Fig. 12L, M). Length up to 5 mm.

Occurs in rather deeper water than *Pseudocuma*. Has been recorded off N.E. England, in the Firth of Forth and Moray Firth, in the Irish Sea and elsewhere from Novaya Zemlya and Spitzbergen southwards to Britain, from the North-West Atlantic southwards to Martha's Vineyard, and also from the Okhotsk and Japan Seas in the North-West Pacific, in depths from about 20 to 2078 m.

Family LAMPROPIDAE

A telson present, of medium or large size, provided with end spines and with the anal opening at its base. Pleopods with an external process on the inner ramus, usually 3 pairs or 0 in the male, occasionally 2 or 1, very exceptionally a single pair present in the female. Exopodites present on the third maxillipeds and the first four pairs of pereopods in the male; in the female on the third maxillipeds and well developed on the first two pairs of pereopods, rudimentary on the third and fourth pairs, or rarely 1 + 1 rudimentary pereopodal exopod on either side or on the first pair only. Inner ramus of uropod with three segments.

Key to the British Species of Lampropidae

1. Telson with 5 apical spines (Fig. 13C). Male without pleopods and with antenna 2 short . . . . *Lamprops fasciata* Sars (p. 44)

Telson with 8 apical spines (Fig. 13G). Male with 3 pairs of pleopods and with antennal flagellum as long as body

Hemilamprops rosea (Norman) (p. 44)

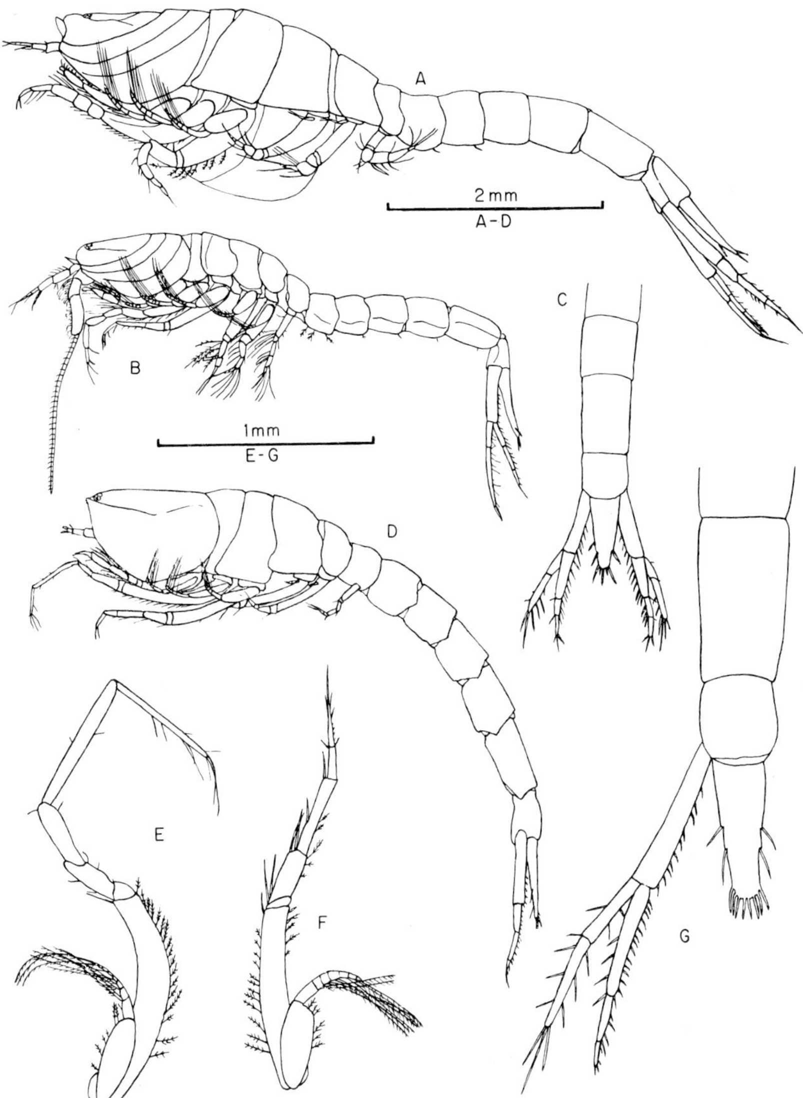

FIG. 13. *Lamprops fasciata:* A, female; B, adult male; C, end of female pleon and uropods. *Hemilamprops rosea:* D, female; E, pereopod 1; F, pereopod 2; G, end of female pleon and left uropod.

Genus LAMPROPS Sars, 1863

Telson large, with more than two apical spines. Carapace with antennal notch fairly well excavated. Antenna 2 of male with flagellum short, prehensile. Pereopod 1 with basis not much shorter than rest of appendage. Male without pleopods. A single British species.

Lamprops fasciata Sars, 1863

Carapace short with three oblique folds on either side. Pseudorostrum short and blunt. Eyes well developed. Pereonites 2 and 3 long (Fig. 13A, B). Male antenna 2 about half as long as the body. Telson rapidly tapering in female with 5 apical spines of which the outer pair and the median are much larger than the intermediate pair, and one or two slender spines on either side (Fig. 13C). Pereopods 3 and 4 of female with rudimentary exopods two-segmented but fairly prominent. In life a broad band of dark violet-brown is present on the carapace with narrower bands on the pereon and pleon. Length of female up to 9 mm, of male up to 7 mm.

A Boreal species found on sand in shallow water all round the British Isles and northwards to northern Norway, from mid-tide level to 70 m depth.

Genus HEMILAMPROPS Sars, 1883

Telson large with more than two apical spines. Carapace of female with antennal notch little excavated. Antenna 2 of male with flagellum reaching to end of pleon. Pereopod 1 with basis much shorter than rest of appendage. Male with 3 pairs of pleopods. A single British species.

Hemilamprops rosea (Norman, 1863)

Carapace short, smooth, with dorsal outline almost straight. Pseudorostrum very short. Second and third pereonites long (Fig. 13D). Telson (Fig. 13G) rather broad, flattened, about as long as the uropod peduncles, not sharply tapered, with 8 spines on the rounded apex and usually two pairs of thin lateral spines. Eyes large, rounded with lenses. Propodus of pereopod 1 slender, much longer than carpus (Fig. 13E). Propodus of pereopod 2 shorter than dactyl (Fig. 13F). Exopods of pereopods 3 and 4 in the female small, two-segmented. In the male the ischium of pereopod 3 carries two falciform spines. When alive the body is covered with crimson chromatophores. Lengths up to 7 mm.

A Boreal species found from North-East England and the Irish Sea to northern Norway in depths from about 5 to 365 m.

Family DIASTYLIDAE

Telson present, usually of medium or large size, less frequently small, usually with two apical spines, sometimes with none. Pleopods without external process on inner ramus, usually two pairs in the male, sometimes rather small, occasionally none. Exopodites present on the third maxillipeds and the first four pairs of pereopods but sometimes only the first two pairs in the male; in the female normally on the third maxilliped but exceptionally absent from this appendage, and on the first two pairs of pereopods, sometimes also as rudiments on the third and fourth pairs, rarely none. Mandibles usually of normal shape, sometimes broadened at base. Inner ramus of uropods with 3, occasionally 2 and rarely 1 segment. Many members of this family show considerable sexual dimorphism.

Key to the British Species of Diastylidae

1. Telson short, not more than half as long as the peduncle of the uropod
   (Fig. 20E) **2**
   Telson distinctly more than half as long as the peduncle of the uropod **4**

2. Telson, excluding spines, about half as long as the peduncle of the uropod **3**
   Telson less than half as long as the peduncle of the uropod (Fig. 20K)
   Leptostylis villosa Sars (p. 58)

3. Telson with only one lateral spine on either side in addition to the terminal
   spines (Fig. 20E) . . . . . *Leptostylis ampullacea* (Lilljeborg) (p. 58)
   Telson with several spines on either side (Fig. 15I)
   Diastyloides serrata (Sars) (p. 48)

4. Pleopods absent or when present without feathered setae (Fig. 15A) (females
   and immature males) **5**
   Pleopods present and with long feathered setae (Fig. 15B) (adult males) **13**

5. Carapace with two or more distinct folds on either side (Fig. 15A) . . **6**
   Carapace without distinct folds (Fig. 15E) **7**

6. Carapace with vertical folds and two pairs of strong dorsal spines (Fig. 19D)
   Diastylis rugosa Sars (p. 56)
   Carapace with diagonal folds and without strong spines (Fig. 15A)
   Diastyloides biplicata (Sars) (p. 48)

7. Pseudorostrum strongly upturned (Fig. 14A)
   Brachydiastylis resima (Kröyer) (p. 47)
   Pseudorostrum not strongly upturned **8**

8. The preanal part of the telson about as long as the postanal part (Fig. 18D)
   Diastylis tumida (Lilljeborg) (p. 54)
   The preanal part of the telson distinctly shorter than the postanal part **9**

9. Telson with not more than 4 lateral spines on either side (Fig. 18I)
   Diastylis lucifera (Kröyer) (p. 54)
   Telson with at least 6 lateral spines on either side **10**

10. A number of large teeth on the front of the carapace, of which one pair are
    very prominent (Fig. 19A) . . . *Diastylis cornuta* (Boeck) (p. 56)
    Only very small teeth or none on the carapace **11**

11. Two long rows of small teeth on the frontal lobe (Fig. 16B)

Diastylis rathkei (Kröyer) (p. 50)

Where there are teeth on the frontal lobe they are arranged transversely **12**

12. Prolongation of the hind end of pereonite 5 at first broad, then ending in a short point. Propodus of pereopod 1 about twice as long as dactyl (Fig. 17C, D) . . . . *Diastylis laevis* Norman (p. 52)
Prolongation of hind end of pereonite 5 ending in a long sharp point. Propodus of pereopod 1 much less than twice as long as dactyl (Fig. 17H, I)

Diastylis bradyi Norman (p. 52)

13. No lateral ridges on the carapace (Fig. 18F) **14**
Lateral ridges, which may be only faintly defined, on carapace (Fig. 16C) **16**

14. Anterolateral angle acute, without serrations below (Fig. 14B)

Brachydiastylis resima (p. 54)

Anterolateral border of carapace rounded, with serrations (Fig. 18F) . **15**

15. Basal segment of endopod of uropod about half as long as remainder

Diastylis lucifera (p. 54)

Basal segment of endopod of uropod longer than remainder

Diastyloides esrrata (p. 54)

16. No vertical or diagonal nor pseudorostral ridges on the carapace (Fig. 18A) **17**
Vertical or diagonal ridges present on carapace, or where absent at least pseudorostral lines present (Fig. 17F) **18**

17. Telson with not more than 4 lateral spines . . . *Diastylis lucifera* (p. 54)
Telson with more than 4 lateral spines . . . *Diastylis tumida* (p. 54)

18. 2 pairs of protuberances on the carapace (Fig. 19B) . *Diastylis cornuta* (p. 56)
No protuberances on carapace **19**

19. Carapace with two narrow ridges running obliquely backwards and not reaching near to the frontal lobe (Fig. 15B) *Diastyloides biplicata* (p. 48)
Carapace without oblique ridges or where present they reach at least to the hind end of the frontal lobe **20**

20. The very distinct pseudorostral lines form an angle broadly open towards the underside (Fig. 17B) . . . . *Diastylis laevis* (p. 52)
The pseudorostral lines are regularly curved or absent (Fig. 17G) . . **21**

21. Pereonites 2–4 depressed in the mid-line so that two distinct but low keels are formed . . . . *Diastylis rathkei* (p. 50)
Pereonites 2–4 without depressions in the mid-line **22**

22. No pseudorostral lines but 2–3 vertical folds on carapace (Fig. 19E)

Diastylis rugosa (p. 56)

Pseudorostral lines present (Fig. 17G) . . . . *Diastylis bradyi* (p. 52)

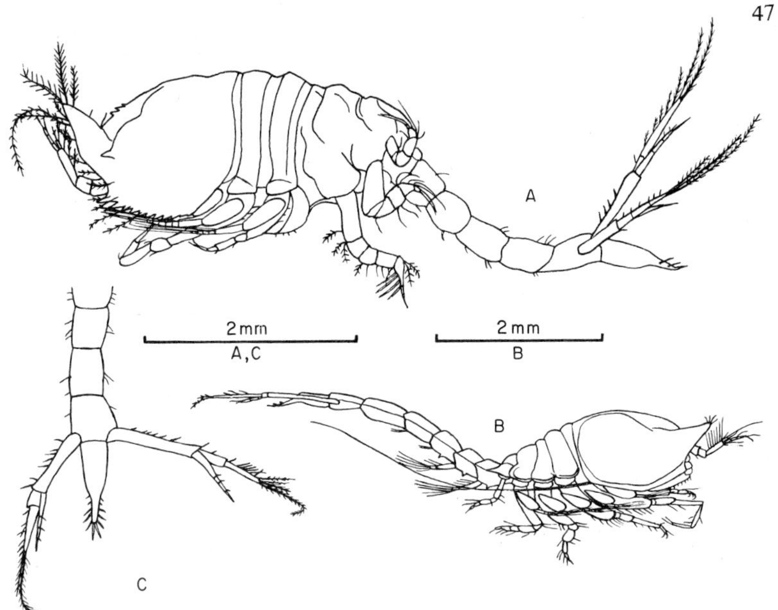

FIG. 14. *Brachydiastylis resima:* A, female; B, adult male; C, end of female pleon and uropods.

Genus BRACHYDIASTYLIS Stebbing, 1912

The pseudorostrum upturned in the female and with several long plumose setae radiating from its end. Carapace short. Female pereonites 3 and 4 coalesced and pereopods 2 and 3 widely separated. Antenna 1 with third segment the longest, bearing a row of plumose setae. Exopod of uropod much longer than endopod. Only one species has been found in British waters.

Brachydiastylis resima (Kröyer, 1846)

Diastylopsis resima Sars, 1900

Female pseudorostrum long and abruptly upturned, with several long plumose setae set around the apex. Anterolateral angle forming a sharp tooth followed by a smaller one and some serrations. Eyelobe barely developed. Frontal lobe with three pairs of denticles (Fig. 14A). Male pseudorostrum little upturned, with a few small setae at the apex, carapace smooth, anterolateral angle rectangular, without teeth or serrations behind (Fig. 14B). Telson about two-thirds as long as the uropod peduncles in the female (Fig. 14C), nearly as long in the male, with about 4 pairs of lateral spines. Endopod of uropod a little more than two-thirds as long as the exopod. Length up to 6 mm.

An Arctic-Boreal species mainly of the upper shelf, recorded from the North Sea northwards to the Arctic and from the north-western Pacific in depths of about 5 to 350 m.

Genus DIASTYLOIDES Sars, 1900

Mandibles truncate at the base, with a small conical projection at the base of the molar process. Rami of the uropods usually nearly equal in length and proximal segment of the endopod longer than the other two segments combined. Male pleopods reduced, with rami one-segmented, the second pair with only one ramus. There are two British species.

Diastyloides serrata (Sars, 1865)

Pseudorostrum fairly well produced, acute, with tip slightly depressed in female. Carapace smooth, with a horizontal slight ridge in its lower half in the male, covered with minute denticles on the branchial regions in the female. Eyelobe very small. Pereonite 5 with posterolateral corners not much produced in the female, acute in the male (Fig. 15E). Telson much shorter than peduncles of uropods, with 5 pairs of lateral spines on the distal half in the female (Fig. 15I), longer and narrower with an angular projection in the male. Peduncle of uropod longer than the rami which are subequal. Length up to 7 mm.

A Mediterranean-Boreal species of the shelf and slope, recorded from Norway to the Mediterranean in depths from about 5 to 1100 m.

Diastyloides biplicata (Sars, 1865)

Pseudorostrum well produced, slightly upturned, especially in female. Carapace with two pairs of lateral oblique carinae and hind margins also carinate (Fig. 15A, B). Telson not much shorter than peduncle of uropods, with 3 to 6 pairs of lateral spines. Peduncles of uropods in female not much longer than the endopod, which is somewhat longer than the exopod (Fig. 15D). Length up to 8 mm.

A Boreal species of the shelf and slope, it has been recorded from Norway to the British Isles in depths from about 60 to 3000 m.

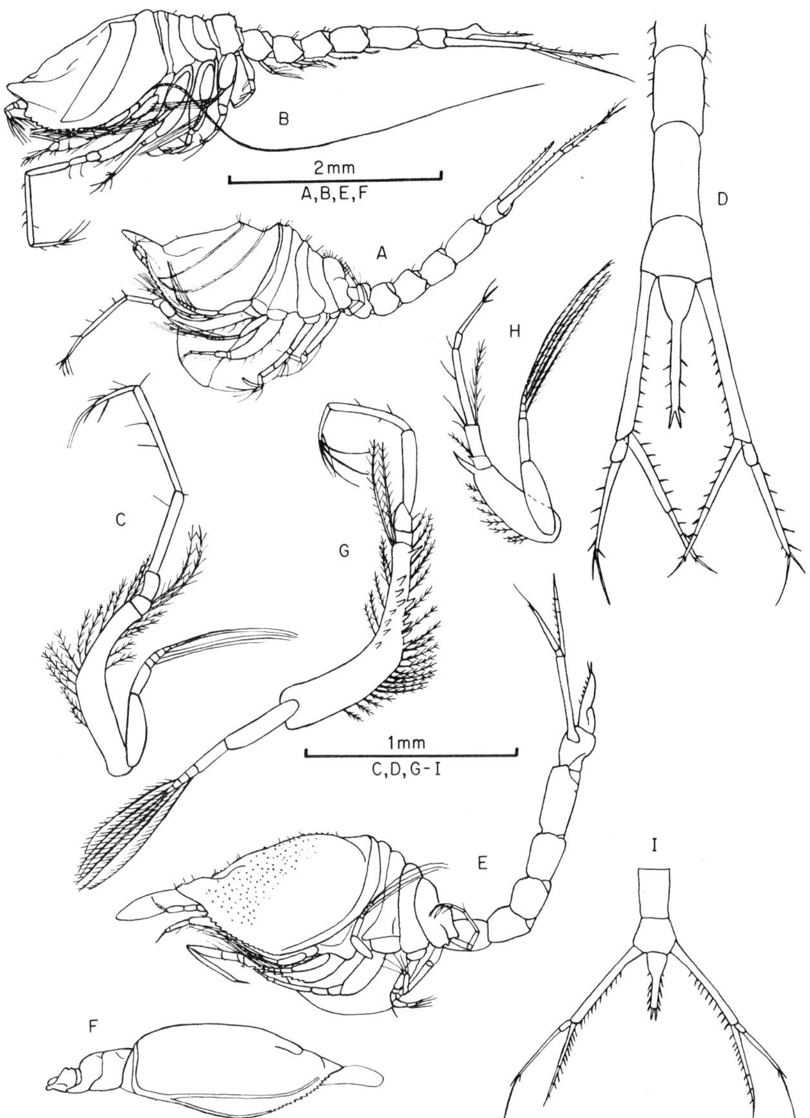

FIG. 15. *Diastyloides biplicata*: A, female; B, adult male; C, pereopod 1; D, end of female pleon and uropods. *Diastyloides serrata*: E, female; F, male carapace and pereon; G, pereopod 1; H, pereopod 2; I, end of female pleon and uropods.

Genus DIASTYLIS Say, 1818

Mandibles normal, not truncated at base. Telson long, postanal part narrowed, with several pairs of lateral spines. Male pleopods well developed, both pairs with two-segmented endopods. A very large genus with seven British representatives.

Diastylis rathkei (Kröyer, 1841)

Pseudorostrum horizontal, conical. Female carapace with a double row of forward-pointing denticles down the middle of the frontal lobe and usually some scattered denticles on the sides, sometimes forming imperfect transverse rows across the frontal area. Eye small but with lenses. Pereonite 5 with posterolateral corners acute, strongly produced backwards (Fig. A, 16B). Telson about as long as or a little longer than the uropod peduncles with 10–15 or sometimes more pairs of lateral spines (Fig. 16F). Pereopod 1 with basis little longer than rest of appendage, with denticles on the side of its distal part; carpus and propodus about equal in length and a little longer than the dactyl (Fig. 16D). Pereopod 2 with dactyl not much longer than propodus, the carpus much longer than both together (Fig. 16E). Male carapace without denticles except on and near the front of the carinae, which on either side curve to run round parallel to the lower margin (Fig. 16C). Length up to 22 mm but usually smaller.

The subspecies *typica* occurs on the coasts of North-West Europe from the southern North Sea and the Irish Sea to Norway in depths of about 10 to 250 m.

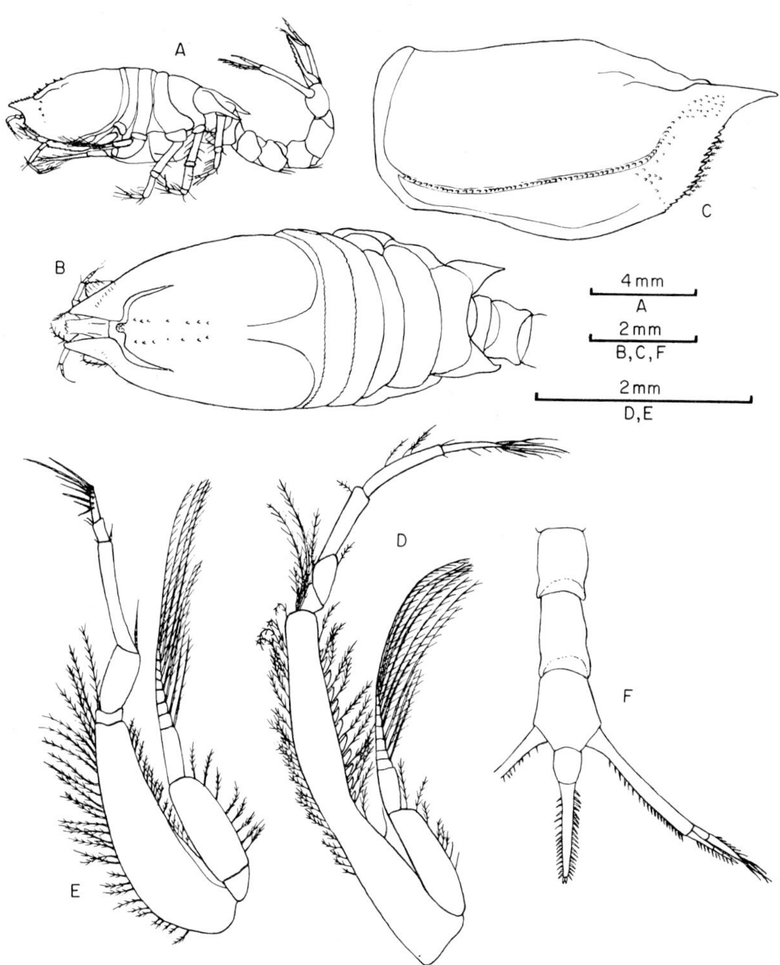

Fig. 16. *Diastylis rathkei:* A, female; B, female carapace and pereon from above;
C, adult male carapace; D, pereopod 1; E, pereopod 2; F, end of female pleon and
right uropod.

Diastylis laevis Norman, 1869

D. rostrata Sars, 1900

Generally resembling *D. rathkei* but carapace of female smooth, with scattered hairs, of male with a smooth longitudinal carina on either side of which the front part forms an obtuse angle in its middle open towards the base and with pseudorostrum pointed, carrying two longitudinal series of spines decreasing in size from front to rear. Eye well developed (Fig. 17A, B). Pereonite 5 of female serrated in front and deeply notched behind, with the lateral angles prolonged into a relatively short point (Fig. 17C), more produced in the male. Telson a little shorter than the uropod peduncles, its distal part, with 14–16 lateral spines, nearly twice as long as the proximal part (Fig. 17E). Basis of pereopod 1 shorter than the ischium to propodus together, the propodus at least as long as the merus and carpus together and about twice as long as the dactyl (Fig. 17D). Length up to 11 mm.

Recorded from the Skagerak to the Ivory Coast, including the British Isles, but not in the Mediterranean, on muddy sand or mud between about 10 and 3000 m but usually on the shelf. According to Lagardère (1971) the only reliable character separating this species from the next is the much longer propodus of pereopod 1 for any given carapace length.

Diastylis bradyi Norman, 1879

Very near *D. laevis* but carapace of female with lines of small denticles, chiefly transverse, the male with the pseudorostral line regularly curved and the pseudorostrum acute, with only a single pair of spines apically or none (Fig. 17F, G). The posterolateral corners of pereonite 5 with much longer acute prolongations (Fig. 17H). There is a sternal spine medially on the female pereonite 3. The basis of pereopod 1 is as long as the remaining segments together and the propodus is much less than twice as long as the dactyl (Fig. 17I). Length up to 12 mm.

This species overlaps the range of *D. laevis* but extends less far to the south. It has been found from the Skagerak to the Bay of Biscay and has probably been confused with *D. laevis*. It occurs on coarser deposits than the latter, usually in shallow water between 0 and 30m depth.

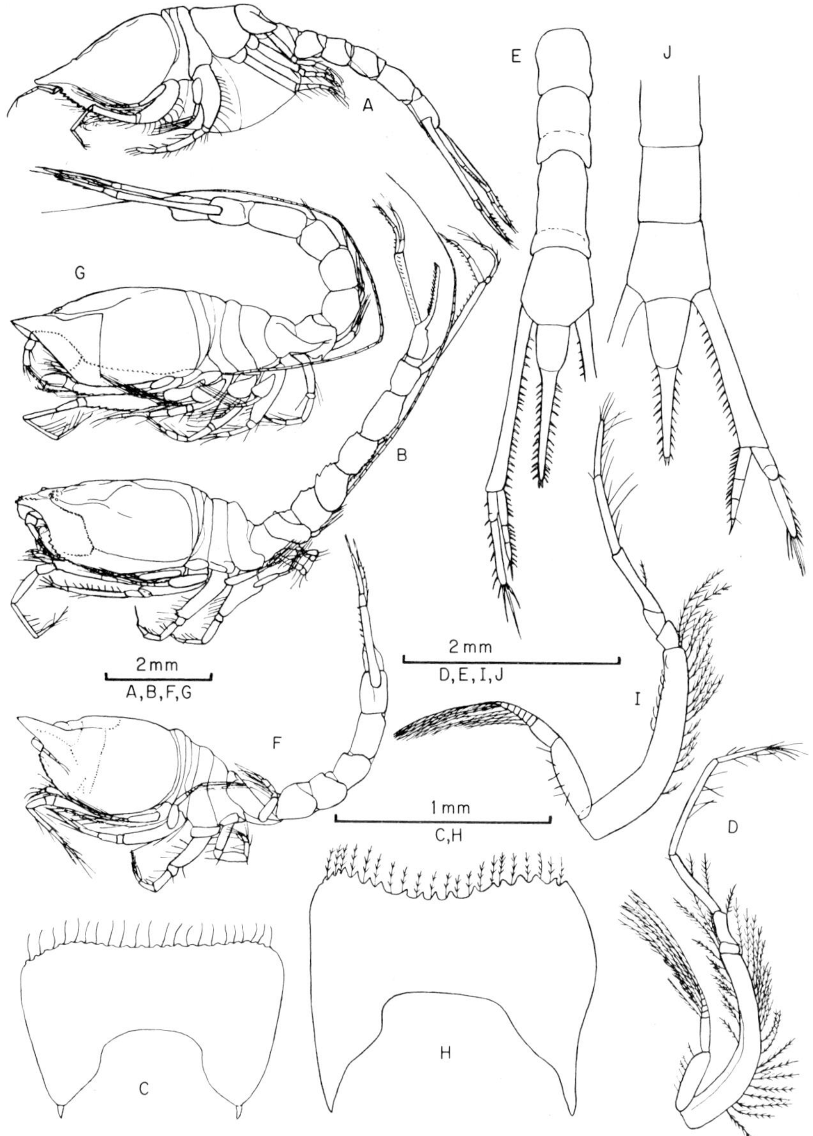

FIG. 17. *Diastylis laevis:* A, female; B, adult male; C, female pereonite 5 from above; D, pereopod 1; E, end of female pleon and left uropod. *Diastylis bradyi:* F, female; G, adult male; H, female pereonite 5 from above; I, pereopod 1; J, end of female pleon and right uropod.

Diastylis tumida (Lilljeborg, 1855)

Carapace as broad as long, smooth in the female or with some minute denticles or more usually with a pair of small teeth one behind the other on either side of the frontal lobe; in the male the lateral line runs near to the lower margin from the subrostral angle to the rear of the carapace. The pseudorostrum is not very long (Fig. 18A). The posterolateral corners of pereonite 5 are rounded in the female, produced to an acute point in the male. The eye is prominent, with lenses. Basis of pereopod 1 with a row of spines, shorter than remaining segments together; the carpus is shorter than the propodus, which is about twice as long as the dactyl (Fig. 18B). The carpus of pereopod 2 is nearly as long as the propodus and dactyl together (Fig. 18C). The telson is shorter than the uropod peduncles and has 8 or 9 pairs of lateral spines; its distal part is about as long as or a little shorter than the proximal part (Fig. 18D). Length up to 10 mm.

Has been collected between northern Norway and the Bay of Biscay and from the Azores, including a few records from N.E. England, the Clyde area and Liverpool Bay off the British Isles, between about 25 and 1400 m.

Diastylis lucifera (Kröyer, 1841)

Carapace of female with denticles in transverse rows on the frontal lobe and less densely behind; in the male only two or three denticles, two flanking the frontal lobe and sometimes one, as in the female, on the eyelobe. Eye small but with lenses (Fig. 18E, F). Pereonite 5 with posterolateral corners little produced in female and not greatly in male. Telson in the female distinctly shorter than the uropod peduncles, with 3 or 4 pairs of lateral spines (Fig. 18I), longer and narrower in the male. Basis of pereopod 1 shorter than rest of appendage, the carpus, propodus and dactyl nearly equal to each other (Fig. 18G). Pereopod 2 with carpus a little longer than the dactyl, which is distinctly longer than the propodus (Fig. 18H). Length up to 8 mm.

A Boreal species found from the Barents Sea to the English Channel, mainly on the shelf, recorded between about 15 and 800 m. It also occurs in the western Atlantic off southern Newfoundland and Maine.

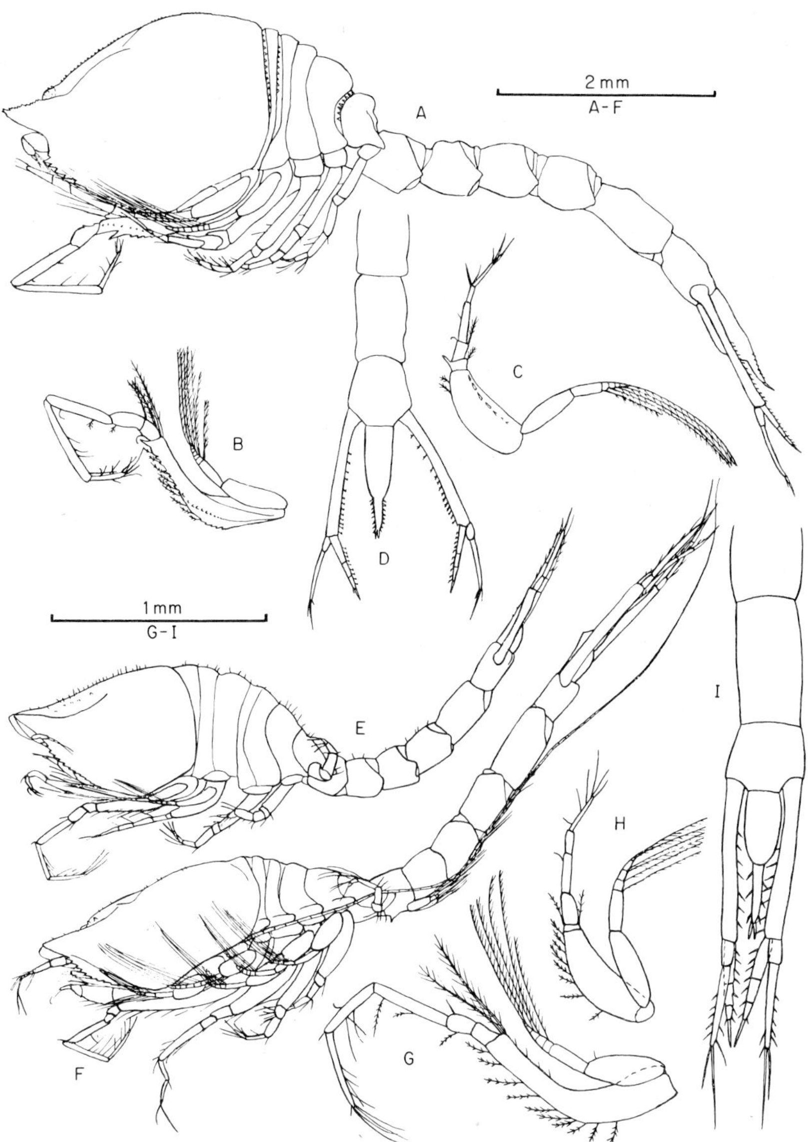

FIG. 18. *Diastylis tumida:* A, female; B, pereopod 1; C, pereopod 2; D, end of female pleon and uropods. *Diastylis lucifera:* E, female; F, adult male; G, pereopod 1; H, pereopod 2; I, end of female pleon and uropods.

Diastylis rugosa Sars, 1865

Female carapace with three to five pairs of not very strongly defined transverse folds, with a strong acute tooth on either side of the frontal lobe and another smaller tooth behind. In the male there are only two pairs of transverse folds but there is a smooth longitudinal carina on either side nearly parallel with the lower edge (Fig. 19D, E). Eye well developed. Posterolateral corners of pereonite 5 rounded in the female, produced to an acute point in the male. Telson a little shorter than the uropod peduncles, its distal portion longer than the proximal and with 8–9 pairs of lateral spines (Fig. 19F). Length up to 9 mm.

A Mediterranean-Boreal species of the upper shelf, found on muddy sand mixed with some coarser material between about 0 and 90 m depth from Norway to the Bay of Biscay and in the Mediterranean.

Diastylis cornuta (Boeck, 1864)

D. capreensis Calman, 1906
D. processifera and *D. stebbingi* Stephensen, 1915

Female with numerous strong spines of which one or two pairs are specially large (Fig. 19A). In the adult male these spines are reduced to two pairs of blunt tubercles, the lower one on either side at the front of a lateral carina, from which an oblique carina extends forwards to the front lower edge of the carapace (Fig. 19B). The eye is small. The telson is a little shorter than the peduncle of the uropods, with its distal part about as long as the proximal and with 8–10 pairs of lateral spines (Fig. 19C). Length up to 14 mm.

Usually occurs on the slope down to 2700 m from northern Norway to the Mediterranean but has been recorded off N.E. England in 37 m and from Shetland. It has in some places been confused with the very similar species *D. boecki* Zimmer, 1930, but the latter has not been recorded with certainty from British coasts.

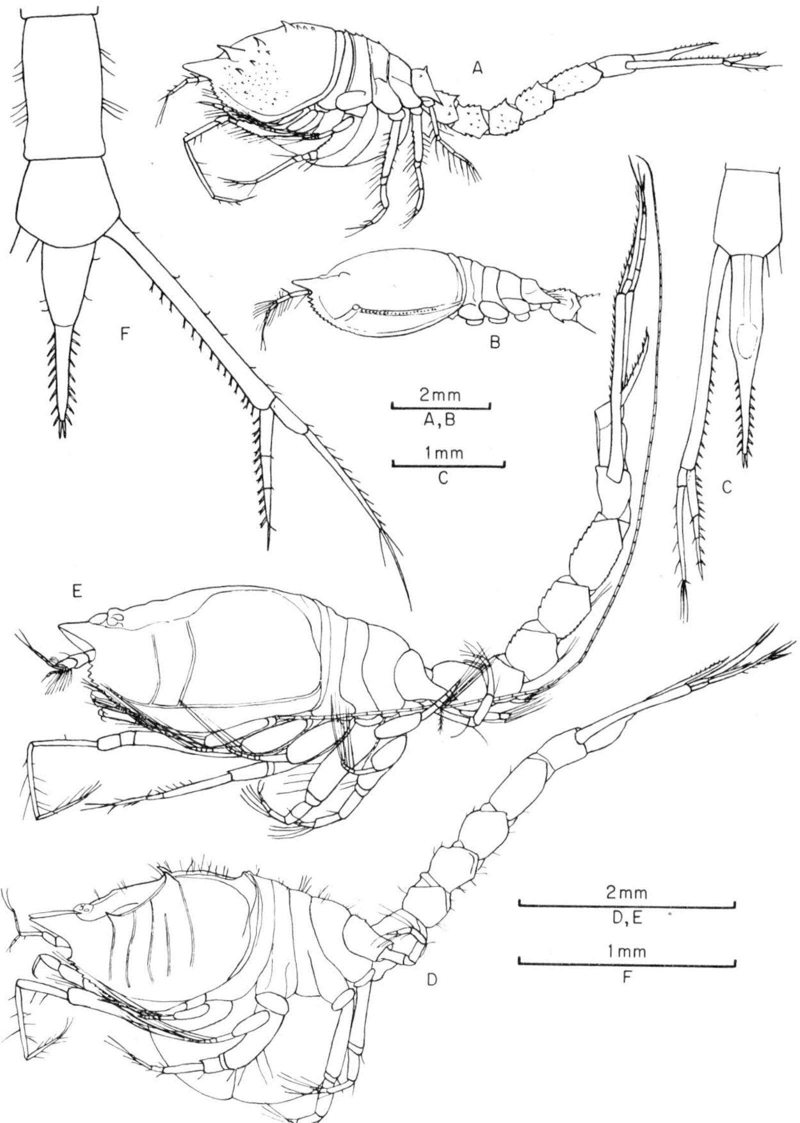

FIG. 19. *Diastylis cornuta*: A, female; B, adult male carapace and pereon; C, end of female pleon and left uropod. *Diastylis rugosa*: D, female; E, adult male; F, end of female pleon and right uropod (A–C after G. O. Sars, 1900).

Genus LEPTOSTYLIS Sars, 1869

Generally similar to *Diastylis* but with a shorter telson having few lateral spines or none. Male antenna 1 with the peduncle dilated, its third segment not the longest and provided with a brush of setae distally. Male antenna 2 shorter than the body. Pereopods 3 and 4 of female with rudimentary exopodites. Exopod of uropods not longer than the endopod. Two species have been found in British waters.

Leptostylis ampullacea (Lilljeborg, 1855)

Pseudorostrum of female short but with a distinct antennal notch below. Carapace broad, with scattered hairs. Anteroventral edges with regular serrations (Fig. 20A, B). Eyelobe very small. Pereonites 1 and 2 of fully adult female produced forwards in a pair of submedian tooth-like processes but not in immature individuals. Telson shorter than pleonite 6, with one pair of small lateral spines (Fig. 20E). Pereopod 1 not very long, with basis much more than half as long as rest of appendage (Fig. 20C). Pereopod 2 with dactyl less than twice as long as the propodus and about as long as the carpus (Fig. 20D). Uropod peduncles about twice as long as the telson, excluding its end spines; endopod with proximal segment nearly as long as second and third combined (Fig. 20E). The adult male has not been described. Length up to 6 mm.

A Boreal species recorded from the North Sea to northern Norway and from the Gulf of Maine, depth about 10 to 550 m.

Leptostylis villosa Sars, 1869

Pseudorostrum of female short, without an antennal notch below. The anteroventral and front edges of the carapace not acutely serrated but cut into square-topped divisions (Fig. 20F, G); the latter, however, are not easily seen. Eyelobe minute. Pereonites 1 and 2 of adult female produced forwards as in *L. ampullacea* but body more densely hairy. Male carapace not hairy, with a machicolated ventrolateral keel on either side; the flagellum of antenna 2 is shorter than the body (Fig. 20H). Telson shorter than pleonite 6 with one pair of lateral spinules (Fig. 20K). Pereopod 1 elongate, with basis not much more than half as long as rest of appendage (Fig. 20I). Pereopod 2 with dactyl much more than twice as long as propodus and distinctly longer than carpus (Fig. 20J). Peduncle of uropods more than twice as long as telson, excluding its end spines; endopod with proximal segment not nearly equal to second and third combined (Fig. 20K). Length up to 6 mm.

Round the British Isles this species has been collected in the North Sea, the Irish Sea and the Firth of Clyde. From there it occurs northwards to northern Norway and possibly into the Arctic, off Iceland and also in the North-West Pacific, in depths from about 10 to 900 m.

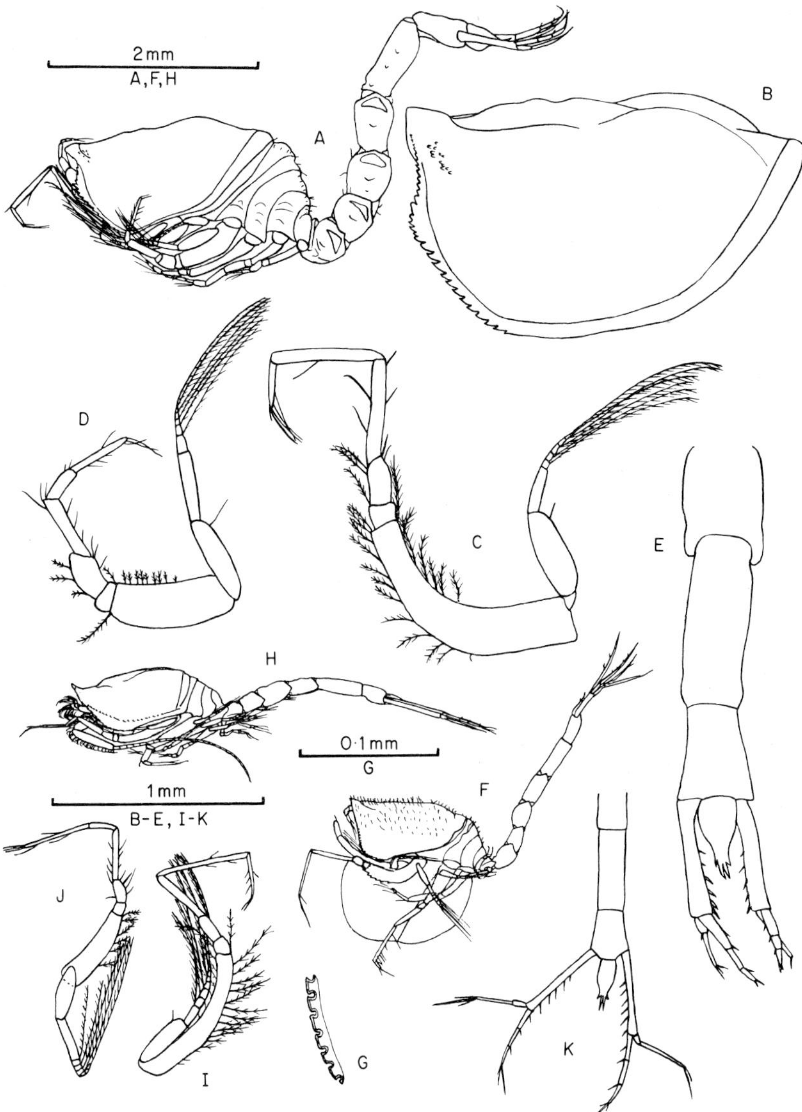

FIG. 20. *Leptostylis ampullacea*: A, female; B, female carapace; C, pereopod 1; D, pereopod 2; E, end of female pleon and uropods. *Leptostylis villosa*: F, female; G, antero-lateral border of carapace in adult female; H, adult male; I, pereopod 1; J, pereopod 2; K, end of female pleon and uropods.

Acknowledgements

I am grateful to Dr. R. J. Lincoln of the Crustacea Section, British Museum (Nat. Hist.) for facilities enabling me to examine and draw a number of the species included in this synopsis.

References

BACESCU, M. 1950. Cumacei Mediteraneeni modificati de mediul pontic. *Anal. Acad. române*, **3** (11), 423–460.

BACESCU, M. 1956. *Cumopsis fagei* n.sp. Cumacé nouveau provenant des eaux du littoral français de la Manche. *Vie et Milieu*, **7**, 357–365.

BACESCU, M. 1972. *Archaeocuma* and *Schizocuma*, new genera of Cumacea from the American tropical waters. *Rev. Biol. Acad. române*, **17** (4), 241–250.

CALMAN, W. T. 1905. The Marine Fauna of the West Coast of Ireland. Part IV. Cumacea. *Sci. Invest. Fish. Brch Ire.* **1904, No.I**, 1–52.

COREY, S. 1969. The comparative life histories of three Cumacea (Crustacea): *Cumopsis goodsiri* (van Beneden), *Iphinoe trispinosa* (Goodsir) and *Pseudocuma longicornis* (Bate). *Can. J. Zool.* **47**, 695–704.

DENNELL, R. 1934. The feeding mechanism of the cumacean crustacean *Diastylis bradyi*. *Trans. R. Soc. Edinb.* **58** (1), 125–142.

DIXON, A. Y. 1944. Notes on certain aspects of the biology of *Cumopsis goodsiri* (van Ben.) and some other cumaceans in relation to their environment. *J. mar. biol. Ass. U.K.* **26**, 61–71.

FAGE, L. 1951. Cumacés. *Faune Fr.* **54**, 1–136.

FORSMAN, B. 1938. Untersuchungen über die Cumaceen des Skageraks. *Zool. Bidr. Upps.* **18**, 1–161.

FOXON, G. E. H. 1936. Notes on the natural history of certain sand-dwelling Cumacea. *Ann. Mag. nat. Hist.* (10), **17**, 377–393.

HESSLER, R. R. and SANDERS, H. L. 1967. Faunal diversity in the deep sea. *Deep-Sea Res.* **14**, 65–78.

JONES, N. S. 1958. Cumacea. *Fich. Ident. Zooplancton*, 71–76.

JONES, N. S. 1963. The Marine Fauna of New Zealand: crustaceans of the Order Cumacea. *N.Z. Dep. sci. industr. Res. Bull.* **152**, 1–82.

KRÜGER, K. 1940. Zur Lebensgeschichte der Cumaceen *Diastylis rathkei* (Kr.) in der westlichen Ostee. *Kieler Meeresforsch.* **3**, 374–402.

LAGARDÈRE, F. 1970. Distinction des cumacés *Diastylis bradyi* Norman et *Diastylis laevis* Norman. *Téthys*, **2** (4), 877–880.

LEDOYER, M. 1965. Sur quelques espèces nouvelles d'*Iphinoe* (Crustacea, Cumacea). Discussion et description comparative des espèces européenes déjà connues. *Rec. Trav. Stn. mar. Endoume*, **39** (55), 253–294.

LOMAKINA, N. B. 1958. Kumovîie raki (Cumacea) Morei SSSR. *Opred. po Faune SSSR*, **66**, 1–302.

MORTENSEN, T. H. 1925. An apparatus for catching the micro-fauna of the sea bottom. *Vidensk. Medd. dansk. naturh. Foren. Kbh.* **80**, 445–451.

NOUVEL, H. 1972. Observations sur les Mysidacés et quelques Cumacés littoraux de la côte française du Golfe de Gascoigne au sud de l'embouchure de la Gironde. *Bull. Cent. Étud. Rech. sci. Biarritz*, **9** (2), 127–140.

OCKELMANN, K. W. 1964. An improved detritus-sledge for collecting meiobenthos. *Ophelia*, **1**, 217–222.

PIKE, R. B. and LE SUEUR, R. F. 1958. The shore zonation of some Jersey Cumacea. *Ann. Mag. nat. Hist.* (13), **1**, 515–523.

SARS, G. O. 1879. Middelhavets Cumaceer. *Arch. Math. Naturvid. Kristiania*, 1878, **3**, 461–512; 1879, **4**, 1–126.

SARS, G. O. 1900. *An account of the Crustacea of Norway*, **3**, Cumacea, 1–115. Bergen.

STEBBING, T. R. R. 1913. Cumacea (Sympoda). *Das Tierreich*, **39**, 1–210. Berlin.

TOULMOND, A. 1964. Inventaire de la faune marine de Roscoff. Cumacés. *Trav. Stn. biol. Roscoff*, **Suppl.** 39–42.

ZIMMER, C. 1941. Cumacea. *Bronn's Cl. Ord. Tierr.* **5**, 1, **4**, 1–222. Leipzig.

Index to Families and Species

For species and genera the correct names are in *italics*; synonyms are in roman. The page citations in roman are to the text; those in *italics* are to illustrations.